Elegant Solutions
Ten Beautiful Experiments in Chemistry

Elegant Solutions
Ten Beautiful Experiments in Chemistry

Philip Ball

Cover image: PhotoDisc

ISBN 0-85404-674-7

A catalogue record for this book is available from the British Library

Published by The Royal Society of Chemistry,
Thomas Graham House, Science Park, Milton Road,
Cambridge CB4 0WF, UK

Registered Charity Number 207890

For further information see our web site at www.rsc.org

Typeset by Alden Bookset, Northampton UK
Printed by TJ International Ltd, Padstow, Cornwall, UK

Contents

Section 2   Posing New Questions

Section 3   The Art of Making Things

Acknowledgements

This book wasn't my idea. Robert Eagling at the Royal Society of Chemistry proposed it to me, and I am grateful for the way he gently yet persistently encouraged me to take the project on. I, of course, bear full responsibility for the lacunae in the final choice of what the book contains. Tim Fishlock at the RSC helped the manuscript through its final stages.

I feel obliged to add a word of warning. I have always tried to ensure that my books can be understood without a scientific training, and this one is no exception to that. But dissecting some of these classic chemistry experiments necessarily takes us deeply into the structures and behaviours of atoms and molecules, and, at the risk of insulting the intelligence of the reader, it is perhaps fair to say that there are a few sections of the book that call for rather more effort from a lay audience than I have asked previously. If you have never before set eyes on a molecular structure, for example, you may find your eyes glazing around pages 169–172, and you should feel no guilt at passing quietly over them.

The immense generosity of several chemists has helped me to assemble the material and to iron out some of the most glaring errors. I am deeply grateful to Jeffrey Bada, Neil Bartlett, Albert Eschenmoser, Leo Paquette, Horst Prinzbach, Matthias Schädel and Claude Wintner for their assistance and comments.

While writing the book, I have had a sense that Oliver Sacks' inspiring enthusiasm for chemistry and its history has somehow been in the air, and in gratitude for his encouragement, support and friendship over the past several years I would like to dedicate this book to him.

Philip Ball
London, February 2005

What is an Experiment? What is Beauty?

Some experiments, it must be said, are best left alone. 'I was desirous', Francis Bacon wrote from his sickbed in 1626, 'to try an experiment or two touching the conservation and induration of bodies; as for the experiment it succeeded excellently well.' An ailing old man of sixty-five, Bacon bought a chicken from a woman in Highgate, a village near London, and filled it full of snow during one of the fierce winters of what we now call the Little Ice Age. This exercise in refrigeration may have worked well enough, but Bacon caught a severe chill and was taken to bed at the house of the Earl of Arundel, where his condition developed into pneumonia. He died within a month.

The irony is that this was one of the few experiments that Bacon, often designated the father of experimental science, actually performed himself. Yet he was a tireless advocate of the experimental method as a way of procuring sound scientific knowledge. 'Our hope of further progress in the sciences', he wrote in his greatest treatise on scientific method, the *Novum Organum*,

> will then only be well founded, when numerous experiments shall be received and collected into natural history, which, though of no use in themselves, assist materially in the discovery of causes and axioms; which experiments we have termed enlightening, to distinguish them from those which are profitable. They possess this wonderful property and nature, that they never deceive or fail you; for being used only to discover the natural cause of some object, whatever be the result, they equally satisfy your aim by deciding the question.

This sounds very much like the traditional formulation of the scientist as someone who devises an experiment to discover something about how the world works, from which more general laws of nature may be deduced. And that is indeed a fair description of the project Bacon envisaged. He rightly complained that, previously, science had lacked any systematic means of gathering reliable knowledge. Instead, it had been pursued (he implied) in a piecemeal fashion by men who sat around thinking up idle dogma based on uncritical acceptance of every report and rumour they heard, or alternatively who carried out 'experiments' with no particular rationale in mind and with little heed to the lessons they might learn. The alchemists, for instance, experimented haphazardly with the sole aim of making gold and getting rich. Thus, said Bacon,

> those who have treated of the sciences have been either empirics or dogmatical. No one has yet been found possessed of sufficient firmness and severity to resolve upon and undertake the task of entirely abolishing common theories and notions, and applying the mind afresh, when thus cleared and levelled, to particular researches; hence our human reasoning is a mere farrago and crude mass made up of a great deal of credulity and accident, and puerile notions it originally contradicted.

And these shortcomings, Bacon felt, existed because no one did decent *experiments*: 'Nothing is rightly inquired into, or verified, noted, weighed, or measured.... We must not only search for, and procure a greater number of experiments, but also introduce a completely different method, order, and progress of continuing and promoting experience.'

This method was what Bacon aimed to set out in his *Novum Organum* or 'New Organon'. The Greek word *organon* means an instrument or engine: Bacon's new engine was the device that would churn out a new philosophical understanding of the world, and it is characteristic of Bacon that he should choose a metaphor from applied science to describe his project. The *Novum Organum* was published in 1620 as a part of a still greater enterprise, *The Great Instauration*, which Bacon intended to be a more or less encyclopaedic account of science as it was then known, coupled with his dream of a new scientific method and a description of the fruits that this approach had already yielded. All Bacon managed to produce of the six volumes intended for his magnus opus, before he was laid low by a frozen chicken, was the introductory material (also published in 1620 under the umbrella title *The Great Instauration*), the second part (which constituted the *Novum Organum*), and a mere sketch of the third volume, *A Preparative Towards a Natural and Experimental History*.

Yet Bacon's vision of an institutional body dedicated to the systematic pursuit of scientific knowledge provided the template for the Royal Society, granted its charter by Charles II in 1665, which brought together (even if it did not exactly unite) such great scientists of the early Enlightenment as Isaac Newton, Robert Boyle, Robert Hooke and Edmond Halley. It is tempting now to regard Francis Bacon as the progenitor of the modern concept of a scientific experiment, wherein accurate and ingenious instruments and devices are deployed so as to reveal the secret workings of the universe.

While that picture has some validity, it isn't quite what Bacon had in mind. His sights were set on a somewhat different destination – for he did not consider the knowledge garnered from experiments to be the ultimate end in itself. Rather, in his famous formulation, knowledge delivers power. The reason why humankind should seek out scientific knowledge is not simply to *know* it but to *apply* it to achieve mastery over nature. Those who merely wove knowledge into intricate, abstract theories, said Bacon, were like spiders spinning their webs. On the other hand, those who seek worldly profit through blind empirical blundering were like ants which 'only heap up and use their store.' The true scientist, he said, should be like the bee, which 'extracts matter from the flowers of the garden and the field, but works and fashions it by its own efforts'. In Bacon's *New Atlantis* (1627), a vision of a utopian society governed by a cadre of scientist-priests who work in a quasi-mystical research institution called Salomon's House, this knowledge produces some truly wondrous inventions:

> We have some degrees of flying in the air; we have ships and boats for going under water, and brooking of seas; also swimming girdles and supporters. We have divers curious clocks, and other like motions of return, and some perpetual motions.

All of this starts to sound rather less hard-headed and more fantastic than we might expect from a man who is ostensibly banishing old superstitions and drawing up a blueprint for a new and reliable scientific method. But that is not surprising; for as science historian John Henry has argued, Francis Bacon's vision of an experimental science drew not so much on the model we would now associate with scientific Enlightenment rationalism, but on the older tradition of natural magic. Experimental science was born out of this magical legacy – a truth acknowledged in the very title of the definitive, multi-volume survey compiled in the 1940s by American historian Lynn Thorndike, *History of Magic and Experimental Science.*

Until we understand this, we will not truly comprehend the experimental tradition in the sciences – and it will be the harder to see how a book like the present one fits into it. To many scientists, a 'beautiful experiment' is one that is perfectly and elegantly designed to yield an insight about the way the world works. That is indeed one kind of experiment, and we can certainly find some beautiful examples of it. But for Bacon, the notion of 'experiment' never lost touch with its roots in the concept of 'art' or *techne*, the Greek word from which 'technology' is derived. It was about making things; and that had, ever since ancient times, the taint of wizardry about it.

During the Renaissance, 'experiment' was sometimes regarded as a dirty word by the Roman Church. According to Pedro Garcia, the bishop of Ussellus in Sardinia, experiments were a form of diabolical magic and should be suppressed:

> To assert that such experimental knowledge is science or a part of natural science is ridiculous, wherefore such magicians are called experimenters rather than scientists. Besides magic, according to those of that opinion, is practical knowledge, whereas natural science in itself and all its parts is purely speculative knowledge.

This was not just because those who dabbled in experiments were liable to be alchemists, astrologers and other heretics who sought to understand and control the occult, demonic forces of nature. It was also because to conduct an experiment was to perpetrate the abominable impiety of asking God a direct question, and perhaps even of coercing him to give an answer. For the thirteenth-century French philosopher William of Auvergne, experimental magic was a 'passion for knowing unnecessary things'. It was in such an intellectual climate that the 'curiosity' that motivated experiments became considered a sin, a 'lust of the eyes' in St Augustine's words.

That is precisely why true science could not even exist without experimentation. The Classical Greek philosophers such as Aristotle and Plato were scientists to the extent that they believed things happened because of natural mechanisms, not through the whimsy of the gods; and they were determined to root out these causes using reason and logic. But on the whole, theirs was the logic of abstract thought, which can never get you very far – because our intuitions about nature are seldom reliable. (You need only read Plato's recommendations for making colours by mixing to realise that he probably never picked up a paint brush.) It wasn't until Greek thought mingled with Middle Eastern artistry in the crucible of polyglot Alexandria that proto-scientists came to appreciate the value of experiment. This philosophical melting pot of Hellenistic Greece produced some of the finest ancient experimentalists, such as Hero and Archimedes.

We have to be careful, however, what we understand by 'experiment' here. Today scientists use a well-designed experiment to probe and perhaps to falsify a theory, or to enable them to choose between different theoretical interpretations. Yet, until the Renaissance, it was extremely rare that an experiment would be conducted to *test* an idea: it was simply a way of demonstrating that you were right.

Even so, there was a difference between actually doing the experiment and just talking about it. The Arabic alchemists of the ninth and tenth centuries appreciated that experimental science must inevitably be a quantitative science. In contrast to the qualitative theories of Aristotle, they gathered knowledge by weighing and measuring, using sophisticated instrumentation: balances, rulers and so on. Experiment creates a demand for instruments, but by the same token, instruments make new experiments inevitable.

So the fact is that, for the Whiggish historian of science (and many practising scientists fall into this category), experiment has roots every bit as disreputable as Garcia implies. Experimental science was not a part of the natural philosophy studied in the medieval and Renaissance universities; it was something done by alchemists and magicians, by the mystical adepts of the Neoplatonic tradition. Medical doctors studied anatomy from books, and the cutting up of bodies was left to unlettered surgeons (which was why the mistakes of the Classical writers persisted for so long). Useful materials like dyes, alkalis and soap were manufactured by artisans and tradesmen. Scholars who, like the thirteenth-century Franciscan monk Roger Bacon, showed an appetite for experiment were inevitably labelled as wizards.

Roger's namesake Francis, nearly four centuries later, professed contempt for alchemists, magicians and their ilk. They had, he said in the *Novum Organum*, so far exerted 'faint efforts' that had met with 'meagre success.' But that was not because natural magic itself was a pile of nonsense; rather, it was because its proponents were hitherto mostly fools and charlatans. Henry points out that Bacon, like most of his contemporaries, 'willingly accepted [that] astrology, natural magic and alchemy were noble and worthwhile pursuits even though, in practice, they were full of error and futility'.

Well, so what? Why should it matter that the 'father of experimental science' drew inspiration from an occult tradition that today plays no role in science? Why should we care that the very concept of a scientific experiment has roots in magic and practical 'arts'?

I believe that bearing this in mind should help prevent us from being too narrow-minded about what we imagine an 'experiment' is. To define it as an enquiry into nature would be to impose a modern definition that denies a great deal of the genealogy of experimental science. I would argue that, at all times before the twentieth century, experimentation was closely linked to *techne*, to applied science and to the skills of the fabricator and the artisan. This perspective, moreover, is particularly important within the context of chemical science, because that discipline has a history that is in many ways quite distinct from the history of physics or biology (with their origins in natural history and the observation of nature). Some areas of what we would now deem to be the sciences of the material world, such as metallurgy, have only rather recently established firm connections with the 'fundamental' sciences on which we now consider them to be based. Likewise, there has been a convergence between some strands of applied chemistry – the manufacture of dyes and pigments, and of soaps and detergents, and the brewing of beverages – with 'academic' science only since the nineteenth century, and that was itself largely driven by the demands of industry for more reliable and versatile methods of synthesis, rather than because academia decided for itself that these 'arts' were worthy of intellectual effort.

This is perhaps why chemistry is so conspicuously absent from some recent books both about the history of science and about its future prospects: it does not 'fit' today's modish model of what science is. The truly bizarre result is that we now have an image of 'science' that is largely at odds with the way it is actually practised. Philosopher of science Joachim Schummer has estimated that there are more – many more – scientific papers published in chemistry than in any other scientific discipline. 'Thus', he says, if we want to know what our actual sciences are about, we should – from a quantitative point of view – first and foremost turn our attention to chemistry. Or, to put it in different terms, philosophies of the natural sciences that neglect chemistry should arouse our strongest suspicion.

What's more, the overwhelming majority of those papers report the results of experiments. 'Chemistry', says Schummer, 'has always been the laboratory science *per se*,'

> such that still in the 19th century the term 'laboratory' denoted a place for experimental research in which *chemical* operations were performed. The chemical laboratory became the model for all the other laboratory sciences when they replaced 'thought experiments' by real experiments. Although chemistry is no longer the only experimental science, it is by far the biggest one and historically the model for all others. Thus, if we want to know what scientists mean by 'experiment', chemical papers are the right point to start with.

Schummer points out that roughly a third of all scientists worldwide are engaged not in the experimental testing of theories, but in producing (and characterizing) new substances – in chemical synthesis. Chemistry, as the eminent French chemist Marcelin Berthelot recognized, creates its own object: it is not necessarily an inquiry into nature, but sets synthetic goals that are shaped by the considerations of the engineer, in particular by the issues of function and design. Synthetic chemistry has its own aesthetic: the 'unnatural' molecules that chemists try to make, while bounded by practical issues such as stability and synthetic accessibility, are ultimately no less a 'designed' product than motor vehicles or buildings, and as such their structure is not inevitable. This brings an added dimension to the notion of a 'beautiful' experiment in chemistry: the beauty need not lie in the conception or the execution, but in the product.

Thus, it seems to me that any attempt to discuss 'beautiful experiments', not just in chemistry but in the whole of science, becomes a skewed endeavour if it neglects that aspect of experimental science engaged in *techne*, in a tradition allied to the arts and crafts, a tradition of making useful and marvellous things – the tradition, indeed, that Francis Bacon drew upon in setting out his ground-breaking plan for giving science a logical and organized structure.

And just what is beautiful?

Good question. Happily, the places and people and things that we find beautiful are many and varied – which means that the selection of experimental examples in this book can never be more than arbitrary. I was heartened by the fact that after I had drawn up a shortlist for the challenge set by the Royal Society of Chemistry – to identify the ten 'most beautiful' experiments in chemistry – I discovered that the American Chemical Society (ACS) had already conducted the same exercise a year previously, and had come to many of the same conclusions as mine. In late 2002 the ACS canvassed its members to submit proposals for the list, and the shortlist of 25 was then assessed and ranked by a panel of chemists and science historians whose combined authority exposes my own list as the scribblings of a rank amateur. At that point, perhaps, I should have just jettisoned my own efforts and adopted the ACS 'top ten'.

But of course, there were parts of that list with which I didn't agree at all. While encouraged by the coincidences with my own choices, I was also stimulated to defend the differences. I ended up striking off my list one experiment that in fact appeared high in the ACS's top ten. Only in one case was I led, after much reflection, to include an experiment I'd originally neglected.

What struck me most about the ACS list, however, was first how it seemed to conflate 'experiment' with 'discovery' – the now pervasive paradigm for historical and philosophical discussions of scientific work. And, second, I noticed how 'beautiful' was often equated by the panellists with what one of them called 'conceptual simplicity', coupled to the lingering notion that a 'beautiful' experiment ought also to be an important one. Indeed, the editorial article accompanying the list defined beautiful in this instance as 'elegantly simple but significant.'

Elegance and simplicity are surely among the key attributes that entitle an experiment to be labelled beautiful, and some of my selections have been made for that reason. But it is not at all clear that these should be the only, or even the principal, criteria for every selection. In fact, if there was ever any intention of that being so for the ACS list, it was flouted more than once. For example, William Perkin's synthesis of aniline mauve, the first aniline dye, in 1856, which fetched in at number 5 in the final ranking, was as messy and inelegant an experiment as one could imagine: the dye was the initially unpromising residue produced by a wholly misconceived attempt at chemical synthesis (see page 154). But the colour itself was surely beautiful, and to my mind that counts for something – albeit not enough to win a place on my list.

As for the issue of significance: there is no real reason why we should demand that a beautiful experiment also be an important one. In practice, that consideration takes care of itself, however, since inevitably the experiments we tend to record and recall and analyse in sufficient detail to know what really happened are those that made an impact. So all of the examples I have chosen do have some broader significance in chemistry or in science more generally. But they are not chosen specifically for that reason, nor are they in any sense meant to represent milestones in the evolution of chemical thought or practice.

I hope I will have said enough by now to justify the position that regards 'experiment' as implying 'experimental science', which could involve a series of investigations, perhaps even spanning several years. This means, however, that compiling a list inevitably means comparing apples and oranges: how do you weigh a single, neat test of some hypothesis against a conclusion derived from the dedicated accumulation of data over a long period? The former can have the beauty of a dramatic revelation; the beauty of the latter can derive from the construction of a coherent chain of logical argument and deduction. For example, experiment number 3 on the ACS list, the determination by the German chemist Emil Fischer in the early 1890s of the precise three-dimensional structure of the glucose molecule, was, in the words of science historian Peter Ramberg,

> part of a large research project involving several smaller projects on the classification of the natural monosaccharides [sugars] that gradually came together in 1891.... There was therefore no one specific experiment that 'determined' the configuration of glucose. This would be an example perhaps of 'beautiful chemical reasoning', rather than a specific experiment.

The same is true of Antoine Lavoisier's work on the oxidation of metals in the seventeenth century, which led him finally to his oxygen theory of combustion. It was a milestone in chemistry (see page 30), it was ranked number 2 in the ACS list – but I am afraid it seemed just too diffuse an endeavour even for me to regard as a single 'experiment'.

Yet I have tried to take a very loose view of how one should regard both 'beautiful' and 'experiment'. One of the key themes in all of the cases I have chosen is that they are shaped by human attributes: invention, elegance, perseverance, imagination, ingenuity. This has tended to work against the inclusion of experiments (like Perkin's) whose success depended on serendipity: chance discoveries are appealing and entertaining, but I find it hard to see beauty in sheer good fortune. (Admittedly, however, most serendipity is more than that.) In retrospect, I realised that each of the selections I have made can be considered to exemplify a different factor that (without providing an exhaustive list) contributes to the beauty of an experiment, and I have suggested as much in my chapter titles.

In the end, I think there are two key reasons why an exercise like this one could be regarded as rather more than sheer indulgence in the current fad for making lists of 'greats' and 'favourites'. One is that it encourages us to think about just what an experiment is and what role experiments play in the evolution of science. It seems absolutely clear that this role extends well beyond the traditional one of hypothesis-testing. Moreover, in researching the histories of some of these experiments I was made aware of the gap that sometimes exists between the popular notion of how they happened and what they meant, and (as far as it can be discerned at all) the historical reality. Experiments give a concrete framework on which to hang stories about the histories of science – but sometimes those stories come to have a strong element of invention about them, which in itself says something interesting about how we understand both science and history.

The second justification for the exercise is that there is nothing like a list to provoke comment and dissent – and thereby, one might hope, to stimulate debate about how science is practiced and about the goals that it sets for itself. I fully expect to be told how outrageous it is that I have omitted this or that experiment from my choices, or that I have included undeserving candidates. In fact, I look forward to it.

CHAPTER 1

How Does Your Garden Grow?

Van Helmont's Willow Tree and the Beauty of Quantification

Vilvoorde, near Brussels, early 17th century—Jan Baptista van Helmont, a Flemish physician, demonstrates that everything tangible is ultimately made from water, by growing a willow tree in a pot of soil nourished by nothing but pure water. His identification of water as the 'primal substance' is consistent with the Biblical account of Creation and thus supports the Christian basis of van Helmont's 'chemical philosophy'. His ideas, published only after his death, represent the final flourish of a semi-mystical view of chemistry that was shortly to give way to the strictly mechanistic philosophy championed by René Descartes.

Perhaps the first thing school students of chemistry learn is that it is all about weighing things. So many grams of this added to so many grams of that: no wonder it so often seems like cookery.

There is nothing obvious about this need for quantification in the study of matter and its transformations. There is little evidence of it in the philosophies of ancient Greece, which sought to explain nature in terms of vague, qualitative propensities and tendencies, affinities and aversions. For Aristotle, things fell to earth because they possessed a natural 'downward' propensity. Empedocles claimed rather charmingly that the mixing and separation of his four elements to make all the bodies of the world were the result of the forces of 'love' and 'strife'.

This is not to say, of course, that quantification was absent from the ancient world. Of course it wasn't. How can you conduct trade unless you know what you are buying and selling? How can you plan a building without specifying the heights and proportions? Throughout the ancient cultures of the Middle and Near East, the cubit was the standard measure of length: the distance from the point of the elbow to the tip of the middle finger. The dimensions of Solomon's Temple are listed in great detail in the Bible's first Book of Kings: an illustration of how much quantification mattered in the court of ancient Israel. Double-pan balances are depicted in Egyptian wall paintings from around 2000 BC, and precious materials were weighed out in grains and shekels. (Because the number of grains to a shekel varied from one country to another, a merchant in the Mediterranean would have to carry several sets of standard stone weights.)

And artisans knew that if you wanted to make some useful substance by 'art' – which is to say, by chemistry, which was then indistinguishable from alchemy – then you had to get the proportions right. A Mesopotamian recipe for glass, recorded in cuneiform script, specifies that one must heat together 'sixty parts of sand, a hundred and eighty parts of ashes from sea plants [and] five parts chalk'. In Alexandria such prescriptions were collated and recorded in alchemical manuscripts, where they began to take on a new character. No longer content with a purely practical, empirical science of matter, the Alexandrian alchemists sought the kind of unifying principles that Greek philosophy extolled. And so one finds tracts like *Physica et Mystica* (as it was known in later Latin translation) by the Egyptian sage Bolos of Mendes, who flourished around 200 BC, in which the recipes are accompanied by the cryptic comment 'Nature triumphs over nature. Nature rejoices in nature. Nature dominates nature.'

Not all of Hellenistic practical science took on this mystical mantle: Archimedes and Hero conducted ingenious and quantitative experiments without conjoining them to some grand theory of nature. Yet for chemistry, the pragmatic and the numinous remained wedded for centuries. When the Arabic philosophers encountered Alexandrian texts during the Islamic expansion in the seventh century AD, they embraced all aspects of its alchemical philosophy. The writings attributed to the Muslim scholar Jabir ibn Hayyan, which were most probably compiled by various members of the mystical Isma'ili sect in the late ninth and early tenth centuries, expounded the idea that all metals were composed of two fundamental 'principles': sulphur and mercury. These were not intended as replacements for the classical Aristotelian elements – Aristotle's philosophy was revered by the Arabs – but they added another layer to it. 'Philosophical' sulphur and mercury were not the elemental substances we now recognize; rather, they were elusive, ethereal essences, more like properties than materials, which were blended in all seven of the metals that were recognized at that time.

Despite their pseudo-theoretical veneer, the Jabirian writings are relatively clear and straightforward in so far as they provide instructions for preparing chemical substances. The great tenth-century Arabic physician Abu Bakr Muhammad ibn Zakariya al-Razi (Latinized as Rhazes) also offered recipes that were very precise in their quantities and procedures:

> Take two parts of lime that has not been slaked, and one part of yellow sulphur, and digest this with four times [the weight] of pure water until it becomes red. Filter it, and repeat the process until it becomes red. Then collect all the water, and cook it until it is decreased to half, and use it.

This prescription produces the compound calcium polysulphide, which reacts with some metals to change their surface colour – a process that would have seemed to be related to the transmutation of one metal to another, the prime objective of later alchemists.

These quantitative recipes, relying on careful weighing and measuring, were copied and adopted uncritically by Western alchemists and artisans in the early Middle Ages. But alchemy was not respectable science: the scientific syllabus at the universities was largely confined to geometry, astronomy and the mathematics of musical harmony. And so while alchemy propagated quantification and motivated the invention of new apparatus, it was indeed largely a kind of cookery learnt from books, and the measurement it entailed did not become a regular part of scientific enquiry. As often as not, old errors of quantification were simply retained. A medieval recipe for making the bright red pigment vermilion from sulphur and mercury – a transformation of obvious alchemical interest – specifies far too much sulphur, because it is based on the Arab alchemists' theoretical ideas about the 'proper' ratio of these substances rather than on their ideal proportions for an efficient chemical reaction.

Only a bold and extraordinary individual would have realized that one's knowledge of the world could be increased by measuring it. The German cardinal Nicholas of Cusa (1401–1464) was such a man. He is one of the great forgotten heroes of early science, an iconoclast who was prepared to make up his own mind rather than taking all his wisdom from old books. In his book *On Learned Ignorance* (1440) (a title that reflected the penchant of scholars for presenting and then synthesizing opposing hypotheses) he argued, a hundred years before Copernicus, that the earth might not be at the centre of the universe. It is a sphere rotating on its axis, said Nicholas, and is larger than the moon but smaller than the sun. And it moves.

For his investigations into natural philosophy he used fine balances and timing instruments such as sand glasses. He suggested that one might observe the rate at which objects fall by dropping them from a tall tower, and cautioned that in such an experiment one should account for air resistance. This demonstrates not only that Nicholas thought to ask quantitative questions (everyone knew that objects fell to earth, but who worried about how *fast* they fell?) but also that he was able to idealize an experimental test: not just to take its outcome at face value, but to think about factors that might distort the result.

To Nicholas's contemporaries, all manner of natural phenomena, such as the weather, were dictated by the influence of the stars. But he laughed at the astrologers, calling them 'fools with their imaginings', and suggested instead that the weather might be forecast not by charting the motions of the heavens but by testing the air. Just leave a piece of wool exposed to the atmosphere, he said – if wet weather looms, the increased humidity will make the wool damp. And what is more, you can put numbers to that: you can figure out how much more humid the air has become by weighing the wool to measure the moisture.

He also had a bright idea for investigating the mystery of how plants grow. The notion of growth from a seed was a central emblem of the mystical philosophy of Neoplatonism, from which most of the medieval ideas about magic and alchemy sprung. But Nicholas saw that this was a problem that could be addressed by quantitative experiment:

> If a man should put an hundred weight of earth into a great earthen pot, and should take some Herbs, and Seeds, & weigh them, and then plant and sow them in that pot, and then should let them grow here so long, until hee had successively by little and little, gotten an hundred weight of them, hee would finde the earth but very little diminished, when he came to weigh it again, by which he might gather, that all the aforesaid herbs, had their weight from water.

It was a fine suggestion; but the experiment was not carried out for another two hundred years.

The troublesome recluse

Nicholas's heliocentrism did not incite the kind of oppression that was famously suffered by Galileo, who had the misfortune to support the idea in less tolerant times. But Galileo's 'martyrdom' was of a relatively mild sort. Giordano Bruno, another heliocentric rebel, was burnt at the stake in 1600 – not, however, for his scientific views but because of his religious heresies. House arrest, to which Galileo was condemned, might seem trivial in comparison; but there was always the threat that it might turn into something worse.

That was largely why the works of Jan Baptista van Helmont (1579–1644) went unpublished in his lifetime. Confined to Vilvoorde in the duchy of Brabant by order of the Inquisition, he did not want any more trouble with the Church. Van Helmont (Figure 1) was no rebel-rouser – in fact he chose to pursue a remarkably quiet, undemonstrative life, turning down offers for appointment as court physician from several princes. Yet this reticence belied an ambition to fashion a chemical philosophy of startling scope – the last, in fact, of its kind – and, when challenged, he did not mince his words.

Van Helmont studied at the University of Louvain, but he felt that academic qualifications were mere vanities and he turned down the degree he had earned. Despite this independence of mind, he was at first something of a medical traditionalist; it was only after he was cured of an itch by an ointment derived from the chemical medicine of the Swiss iconoclast Paracelsus that he converted to this new kind of 'physick'. Whereas traditional medicine throughout the Renaissance was based on the ideas of the Greek doctor Hippocrates and the Roman Galen, which held that health was governed by four bodily fluids called humours, Paracelsus (1493–1541) maintained that specific diseases should be treated with specific remedies created from nature's pharmacopoeia by the art of alchemy. Several decades after his death, Paracelsus's ideas gained popularity throughout Europe, and by the early seventeenth century the medical community was divided into Galenists and Paracelsians.

Van Helmont studied the writings of Paracelsus and found much there that seemed to him to be sound advice. But he was by no means an uncritical disciple. Paracelsus tended to surround his chemical medicine with a fog of obscure terminology and overblown notions of how the world worked. Humankind, he said, was a microcosm reflected in the macrocosm of the universe, so that the disorders of the body

Figure 1 *Flemish physician and alchemist Jan Baptista van Helmont*
(Reproduced Courtesy of the Library and Information Centre, Royal
Society of Chemistry)

could be compared to the disorders of nature – epilepsy, for example,
known as the falling sickness, was akin to the tremors that shook the
ground in an earthquake. This concept of a correspondence between
the microcosm and the macrocosm was a central theme in Neoplatonic
philosophy and was popular with the Jabirian alchemists. But to van
Helmont it looked like sheer mysticism, and he would have none of it.

Instead, he pursued the difficult task of separating what was worthy
in the works of Paracelsus from what was nonsense: he wanted the
chemical medicine without the chemical philosophy. But that did not

mean he was free of mysticism himself, for like Paracelsus he felt it was essential that chemical science be based in Christian theology. In his own mind he was replacing speculation with rigorous theory; but from today's perspective there is often not a great deal to differentiate the philosophy of Paracelsus from that of van Helmont.

For example, van Helmont supported the Paracelsian cure known as the weapon salve, an idea that seems now to be ridiculously magical. To cure a wound made by a weapon, you should prepare an ointment and then apply it not to the cut but to the blade that made it. However unlikely a remedy, van Helmont was convinced that it had a perfectly rational, mechanistic explanation. The natural magic of the Neoplatonists was not mere superstition; it was based on the belief that the world was filled with occult forces, of which magnetism was an incontestable example. The weapon salve mustered these forces to allow the vital spirits of the blood on the blade to reunite with that in the body.

When van Helmont published a defence of the weapon salve in 1621, it was criticized by a prominent Jesuit. Van Helmont responded by explaining the 'mechanism' of the cure, and he rather unwisely compared it to the way religious relics produce 'healing at a distance'. The University of Louvain found this a scandalous thing to suggest, and van Helmont's ideas were brought before the Spanish Inquisition (Spain ruled the Low Countries at that time). He was declared a heretic, and was lucky to escape with nothing more severe than a spell in prison before being freed through the intervention of influential friends. Thereafter, van Helmont was forbidden to publish anything further without the approval of the Church, or to leave his home without the permission of the Archbishop of Malines – a restriction that applied even in times of plague. During one outbreak, his family refused to leave the town without him, and two of his sons succumbed to the disease.

So his writings on chemistry and medicine were not published until after his death, when his son Franciscus Mercurius inherited his manuscripts. Van Helmont's collected works appeared in Latin under the title *Ortus Medicinae* (Origins of Medicine) in 1648, which John Chandler translated into English in 1662 as *Oriatrike; or, Physick Refined*.

Ortus Medicinae contains a wealth of striking ideas, most notably the suggestion that digestion (which Paracelsus saw as an alchemical process conducted by an 'inner alchemist' called the Archeus) is a kind of fermentation involving an acid. The book is a curious mixture of new and old, prescient and regressive. Just as the mechanistic philosophy of Descartes and his followers was taking hold in Europe (and shortly before it was to be refined in Isaac Newton's *Principia Mathematica*), van Helmont challenged the Cartesian division of body and soul by arguing for a kind of vital force that animated all matter. Van Helmont believed that he would find this 'world spirit', the *spiritus mundi*, by distilling blood.

At the same time, he called for an end to the sort of science that relied solely on logical thinking and mathematical abstraction – it should instead be based on observation, on experiment. As a demonstration of what could be gained that way, van Helmont explained how he had come to understand that everything was made from water.

Well, not quite everything. The other of the Aristotelian elements that he continued to countenance was air. But this air, he said, is inert and unchanging, and so all else is nothing but water. 'All earth, clay, and every body that may be touched, is truly and materially the offspring of water onely, and is reduced again into water by nature and art.'

In support of this claim, van Helmont explained how 'I have learned by this handicraft-operation, that all Vegetables do immediately and materially proceed out of the Element of water onely.' Whether or not he knew of the experiment proposed by Nicholas de Cusa, he had actually gone ahead and done it.

It required the kind of patience that perhaps house arrest cultivates in a person. Van Helmont took 200 pounds of earth, which he dried in a furnace and then moistened with rain water. He placed it in a pot and planted within it a small willow sapling weighing five pounds. And then he waited for five years.

He watered it whenever necessary, but carefully excluded all other sources of matter. Van Helmont explains how, to keep out dust, he 'covered the lip or mouth of the Vessel, with an Iron Plate covered with Tin', which was 'easily passable with many holes' to let through water and air. In other words, like Nicholas de Cusa he was thinking about how to exclude influences that could corrupt his results.

At the end of that time he weighed the tree again, and also the soil, which was only about two ounces short of the original 200 pounds. The tree, however, had grown immensely. 'One hundred and sixty-four pounds of Wood, Barks, and Roots arose out of water onely', he said. And he added that he had not included in this estimate the weight of the leaves that had grown and then fallen over four autumns.

One might argue that the experiment hardly required such quantification. Anyone could see that the soil had not greatly diminished in volume, while the tree had very obviously gained a lot of mass. In any case, what was the significance of the figure of 164 pounds, if the leaves were neglected? But that wasn't the point. Numbers are hard facts; they are irrefutable. If anyone doubted the interpretation, van Helmont could demand that they kindly explain where else one hundred and sixty four pounds of material had come from (and you can imagine how absurd it would have been to suggest that this matter came out of the *air*!).

The experiment was beautiful because of the clarity of its concept: it was hard to see what could possibly have been overlooked, or what could have led to any error. That beauty is enhanced by the reliance on quantification, which transforms an anecdote into a scientific result. All of which makes it perhaps rather shocking that van Helmont was of course completely wrong: wood is not made from water, but from atmospheric carbon dioxide absorbed through the leaves and converted into cellulose by photosynthesis. It is hard to fault either the experimental design or the logic of the interpretation; we can't reasonably expect van Helmont to have come to any other conclusion. There is surely a humbling message in this for scientists today: if an important part of the puzzle is missing, what seems 'obvious' may in fact be fundamentally fallacious.

End of an era

This was not the sole extent of van Helmont's evidence for making water the prime matter of the world. But the rest of his argument was largely circumstantial, and lacked such quantitative exactitude. What else nourishes fish, if not water? Don't solids of all kinds turn into water when they come into contact with it – salts, for example, which produce 'savoury waters' when they dissolve? Of course, there are plenty of solids that do not dissolve, but van Helmont believed this was just because the right solvent hadn't been found (and in certain respects he was right!). He spoke of a 'universal solvent' that would dissolve all things, which he called the *alkahest*, and he spent many fruitful hours searching for it. (It's not clear what, if he had been successful, he proposed to keep it in.)

Equally important was the evidence from Holy Scripture. Was it not made clear in Genesis that God created the world out of water, by separating 'water from water' and placing in the gap first the expanse of the sky and then dry ground? At the dawn of the Age of Enlightenment, theology still carried some weight in matters of science.

Yet he also adduced an ingenious piece of alchemy to support his contention. He could even turn sand into water, by melting it with an alkali to make 'water glass' (sodium silicate), which will liquefy as it absorbs moisture from the air. Add an acid, and the sand is regenerated in precisely the same amount. The quantities were again important here: van Helmont was convinced that matter was indestructible, so that it was conserved in any transformation of this sort.

Van Helmont was not the first person to propose that the world could be built from water alone. The Greek philosopher Thales, founder of the influential Ionian school, said as much in the sixth century BC, and part of his reasoning was similar – for water can be converted into 'air' by evaporation, while freezing transforms it into 'earth', which is, to say, a solid. But Thales' idea never caught on, even among the later Ionian philosophers – and neither did van Helmont's.

There is no compelling scientific reason why this should have been the case; rather, one might say that the circumstances were not to van Helmont's advantage. For one thing, all-embracing 'chemical philosophies' were about to be eclipsed by Cartesian mechanistic science in the mid-seventeenth century: van Helmont's writings represent their final bloom. Although he helped to place Paracelsian science on a more rational basis, he didn't go nearly far enough; men like the Germans Andreas Libavius and Johann Rudolph Glauber were yet more ruthless in stripping chemistry of its Neoplatonic, magical trappings. At the Jardin du Roi, the royal medical and pharmaceutical school in Paris, alchemy was evolving into the academic discipline of 'chymistry'. And the year before van Helmont's *Oriatricke* appeared in England, Robert Boyle published his epoch-making critique of earlier ideas on chemistry, *The Sceptical Chymist*, which warned that chemists should be rigorous about how they defined an element and should not extrapolate beyond what the evidence permitted.

Besides, there were many systems of elements to choose from in the seventeenth century – several of them amalgams of Aristotle's quartet and Paracelsus's alchemical triumvirate of sulphur, mercury and salt – and van Helmont's two-element scheme really did not have much more to recommend it above any other. In addition, it did not help that chemical philosophies had come to be associated with politically radical factions, such as the Bohemian rebels who denied the authority of the Holy Roman Emperor in 1619 and thereby triggered the Thirty Years' War. In England too, Cromwell's Puritans looked askance at such radicalism.

But van Helmont left his mark in other ways. He was interested in the 'spirits' that could be produced in chemical processes such as combustion, which were clearly different from ordinary air. He collected one such vapour, the 'spirit of wood', that was released from burning charcoal, and found that it could extinguish a flame. He was sure that these vapours were derived not from air but from water, and he decided they needed a new name. He borrowed a term that Paracelsus had used, the ancient Greek word *chaos*, which he transliterated as it sounded on the Flemish tongue: 'gas'. What were these gases? That question was to set the principal research agenda of the chemists of the next century.

CHAPTER 2

An Element Compounded

Cavendish's Water and the Beauty of Detail

London, 1781—The eccentric aristocrat Henry Cavendish, one of the wealthiest men in England, ignites two kinds of 'air' in a glass vessel and finds that they combine to form water. It is an experiment that has been performed before, and one that will be repeated subsequently by several other scientists. But Cavendish subjects the process to greater scrutiny than anyone previously, making careful measurements of all the quantities concerned, and his results point the way to a more definitive and remarkable conclusion: that these 'airs' are the very constituents of water, previously considered to be an irreducible element.

But is that what Cavendish himself thought? The issue, and with it Cavendish's claim to the discovery that water is a compound, were hotly contested in the nineteenth century. This 'water controversy' is further clouded by Cavendish's gentlemanly disregard for acclaim, which meant that he did not hurry into print but examined his findings for a further three years before publishing them. In the meantime, others scented the same trail, and the result was a priority dispute that historians are still debating today.

Even though van Helmont's belief in water as the fundamental stuff of all creation was not taken seriously by the late eighteenth century, nonetheless there seemed little reason to doubt that water was an element – the last, perhaps, of the Aristotelian elements to remain unchallenged. The problem is that when everyone believes something, no one bothers to check it. When he performed his famous experiment,

Henry Cavendish was not setting out to investigate the nature of water. Like many of his contemporaries, he was more interested in that other ancient element: air.

This was the age of 'pneumatic chemistry', when researchers devoted themselves to collecting the 'vapours' that bubbled from chemical processes. Once considered inert and therefore uninteresting, 'air' was now found to come in several varieties. The English clergyman Stephen Hales showed in 1727 that 'airs' could be collected by bubbling them through water to 'wash' them, and then collecting them in a submerged, inverted glass vessel. The 'Hales trough' allowed one to quantify the amount of 'air' collected by observing the volume of water it displaced.

The Scotsman Joseph Black used the technique to study an 'air' produced by heating limestone or magnesia: this vapour seemed to be miraculously 'fixed' in the minerals until heat drove it out, and Black called it 'fixed air'. It was not like 'common air': substances wouldn't burn in fixed air, and it had the signature property of turning lime water (a solution of calcium hydroxide, then known as slaked lime) cloudy. And then there was the deathly 'mephitic air' identified by Black's student Daniel Rutherford, a residue of common air that remained after combustion was carried out in a sealed vessel. The 'chymists' of the seventeenth century had known about another vapour produced when acids acted on certain metals: the Swedish apothecary Carl Wilhellm Scheele collected this gas in 1770 and observed that it burnt explosively in common air. Scheele called it 'inflammable air'.

The chemistry of airs had a theory, and it was based around the substance called phlogiston. In 1703 the German chemist Georg Stahl named this mercurial substance after the Greek word *phlogistos*, 'to set on fire'. Phlogiston was what made things burn. Some substance, said eighteenth-century scientists, was being transferred between the air and a combusting material – and that substance was phlogiston.

Materials were considered to lose phlogiston when they burnt.* When common air was saturated with phlogiston, burning ceased: that was why a candle inside a sealed vessel would eventually go out. For the English Nonconformist minister Joseph Priestley, this explained the character of Rutherford's mephitic air: it was nothing but normal air mixed with a sufficiency of phlogiston. In 1774 Priestley discovered

---

* That was why wood got lighter as it was consumed by flames. But, inconveniently, metals got heavier when they were heated (calcined) in air, even though they were supposed to be losing phlogiston. No problem, said the advocates of phlogiston theory: apparently this volatile 'principle of combustion' can sometimes have negative weight.

how to make the opposite of this lifeless, smothering substance: how to create an 'air' that was 'dephlogisticated' and thus wonderfully conducive to combustion. He made it by heating mercury oxide, something that others (including Scheele) had done before.

In the same year, Priestley's friend John Warltire looked carefully at the explosive combustion of Scheele's inflammable air. Warltire seized on the contemporary fad for investigating electricity by using an electrical spark to ignite a mixture of common air and inflammable air, and he found that after the explosion there was less 'air' than before, and that the walls of his vessel were coated with dew. In Paris, Pierre Joseph Macquer found much the same thing: inflammable air burnt with a smokeless flame, and when a porcelain plate was placed over the flame, it was moistened with drops 'which appeared to be nothing else but pure water'.

And so what? Everyone knew that water could condense out of common air to mist a window with droplets or to make the pages of books curl up in dank cellars. Warltire did not much concern himself with the water, and neither did Priestley when he repeated the experiment in 1781. They were more interested in what was happening to the 'airs', and what this meant for phlogiston theory. Inflammable air was clearly rich in phlogiston – indeed, some scientists, including Scheele and Cavendish himself, suspected that it might be pure phlogiston – and Priestley figured that this phlogiston caused common air to release the water it contained: 'common air', he said, 'deposits its moisture by phlogistication'.

Then Henry Cavendish decided to take a look too.

A queer fellow

More than any other science, chemistry has traditionally told its history through a progression of colourful characters. Empedocles, drunk on dreams of immortality, throws himself into Mount Etna; Paracelsus staggers foul-mouthed and drunken through Renaissance Europe; Johann Becher, the wily alchemist who started the whole phlogiston business, swindles the princes of the Nertherlands with promises of alchemical gold; Dmitri Mendeleyev, who drew up the periodic table, is the wild and shaggy prophet of Siberia. Most of these tales contain a strong element of hearsay, if not outright invention. And Cavendish can be relied upon for a gloriously odd comic turn. In Bernard Jaffe's *Crucibles*, the archetype for this kind of history, Cavendish was

THE HONOURABLE HENRY CAVENDISH

Figure 2 *Henry Cavendish: this ink-and-wash sketch by William Alexander, the only known portrait of the reclusive scientist, was prepared by the artist with not a little subterfuge*
(Reproduced Courtesy of the Library and Information Centre, Royal Society of Chemistry)

'gripped by an almost insane interest in the secrets of nature, ... not giving a moment's thought to his health or appearance'. The son of Lord Charles Cavendish and heir to a fortune, he 'never owned but one suit of clothes at a time and continued to dress in the habiliments of a previous century, and shabby ones, to boot' (Figure 2).

If this Cavendish is a stage character, however, there is no denying that he is more than just invention. The Honourable Henry Cavendish was genuinely strange and difficult to know; his own colleagues make that clear enough. Charles Blagden, Cavendish's associate and the only person with whom he seems to have had anything approaching a close relationship, calls him sulky, melancholy, forbidding, odd and dry. The scientist and politician Lord Brougham, 35 years after Cavendish's death, says that he 'uttered fewer words in the course of his life than any man who ever lived to fourscore years, not at all excepting the monks of La Trappe'. He recalls how Cavendish would shuffle quickly from room to room at the Royal Society, occasionally uttering a 'shrill cry' and 'seeming to be annoyed if looked at'.

Even the usually generous Humphry Davy, who said on Cavendish's death that since the demise of Isaac Newton England had suffered 'no scientific loss so great', found the man himself 'cold and selfish' (he made the same charge of Blagden). Davy admitted that Cavendish was 'afraid of strangers, and seemed, when embarrassed, even to articulate with difficulty'. The chemist Thomas Thomson called him 'shy and bashful to a degree bordering on disease'.

That seems indeed to be the true measure of the man. Contrary to what Jaffe suggests, Cavendish may not have been exactly misanthropic but, rather, painfully shy to the point where he was barely able to interact at all with his fellows. If he seemed 'cold', it is likely that this was simply the appearance conveyed by his extreme diffidence. Perhaps the most telling image we have is that of Cavendish hovering on the doorstep of the house of Joseph Banks, the Royal Society's president, unable to bring himself to knock on the door and face the crowds within.

On the basis of the biography of Cavendish published in 1851 by chemist George Wilson, Oliver Sacks has made a tentative diagnosis of the subject's social dysfunction:

> Many of the characteristics that distinguished Cavendish are almost pathognomic of Asperger's syndrome: a striking literalness and directness of mind, extreme single-mindedness, a passion for calculation and quantitative exactitude, unconventional, stubbornly held views, and a disposition to use rigorously exact (rather than figurative) language – even in his rare non-scientific communication – coupled with a virtual incomprehension of social behaviours and human relationships.

There seems to be sufficient consensus among contemporary descriptions of Cavendish's behaviour to make such a conclusion likely.

But Wilson's biography, while often taken at face value, was not a dispassionate account of the man; it had an agenda, as we shall see.

Yet for all his reticence, Cavendish scarcely ever missed the weekly dinner of the Royal Society Club at the Crown and Anchor on the Strand, nor was he often absent from the Monday Club at the George & Vulture coffee house. Although conversation seemed an agony to him, he forced himself into society, because in the end he wanted to mix with his learned colleagues and share with them the adventure of science.

For that was the life Cavendish chose. Like his father, he could have followed the conventional political career of an aristocrat; but like Charles he turned instead to science. He had only just been elected a member of the Royal Society when, in 1766, he published a stunning paper in the society's *Philosophical Translations* (he never published anywhere else) on the chemistry of airs. 'Three Papers, Containing Experiments on Factitious Air' won him the Royal Society's prestigious Copley Medal.

'Factitious' meant any air that was somehow contained within other materials 'in an unelastic state, and is produced from thence by art'. Black's fixed air was such a gas, and inflammable air was another. Cavendish was not content with noting that this latter air went pop when ignited; he reported careful measurements showing that it was 8700 times lighter than water and capable of holding '1/9 its weight of moisture'.

This kind of detail reveals the way Cavendish thought about experiments. His laboratory, housed within the grounds of his ample townhouse in Great Marlborough Street, near Piccadilly in London, was filled with measuring devices. The caricature presented by Wilson, and more or less uncritically repeated ever since, shows Cavendish as a calculating machine, obsessed with quantification; but the fact was that he understood this was now the only reliable way to do science. We've seen that van Helmont recognized the value of measurement in the seventeenth century; but Cavendish's vision penetrated further than that. He understood the meaning of accuracy and precision, and realised that all experiments have a finite and unavoidable margin of error. He estimated the accuracy of his determinations, making distinctions between the errors introduced by the experimenter and the limitations of the instrumentation. To reduce such sources of error, he would repeat experiments and take averages of the results. And he would quote numerical results only to the appropriate number of significant figures. The great French scientist Pierre-Simon Laplace, who pioneered statistical techniques for handling errors in experiment (and of whom more later), remarked to Blagden that Cavendish's work was conducted with the 'precision and finesse that distinguish that excellent physicist'. This is arguably Cavendish's greatest contribution to experimental science: an attention to numerical detail that keeps the experimenters' claims in proportion to what their methods justify.

And numbers have power. By putting numbers on the low weight of this vapour relative to common air, Cavendish excited speculations about whether it might enable a man to 'fly' by means of the buoyancy of a balloon filled with it. And so it did: the physicist Jacques Charles took to the air in 1783 in Paris, prompting Antoine Lavoisier to scale up his method of producing 'inflammable air' while Joseph Banks covered up his nationalistic chagrin with sniffy remarks about the flighty French.

In the early 1780s, Cavendish decided to explore 'the diminution which common air is well known to suffer by all the various ways in which it is phlogisticated'. In other words, he was keen to examine the process that Warltire and Priestley had described, in which common air is reduced in volume by igniting it with inflammable air (which might or might not be phlogiston itself). Thus he was not, in a sense, proposing to do anything new; rather, he saw that sometimes an experiment yields its secrets only when you start to look at the details. 'As the experiment seemed likely to throw great light on the subject I had in view', he explained in the report of his studies, presented to the Royal Society in 1784, 'I thought it well worth examining more closely'.

Anatomy of an explosion

Like the others before him, Cavendish made inflammable air by dissolving zinc or iron with acids, and he set off the detonation with a spark. 'The bulk of the air remaining after the explosion', he wrote,

> is then very little more than four-fifths of the common air employed; so that as common air cannot be reduced to a much less bulk than that by any method of phlogistication, we may safely conclude that when they are mixed in this proportion, and exploded, almost all the inflammable air, and about one-fifth part of the common air, lose their elasticity, and are condensed into the dew which lines the glass.

Every detail was carefully checked out; nothing was taken for granted. The dew, he said 'had no taste nor smell, and ... left no sensible sediment when evaporated to dryness; neither did it yield any pungent smell during the evaporation; in short, it seemed pure water'. In some experiments he noticed that the explosion produced a little 'sooty matter', but he concluded that this was probably a residue from the putty ('luting') with which the glass apparatus was sealed; and indeed 'in another experiment, in which it was contrived so that the luting should not be much heated, scarce any sooty tinge could be perceived'.

Was the dew truly pure water? Cavendish found in some initial experiments that it was in fact slightly acidic, and he spent long hours tracking down where the acid came from. Although he did not put it quite this way himself, the acidity stems from reactions between oxygen in the air and a little of the nitrogen that makes up the 'inert' four-fifths of the remaining gas, creating nitrogen oxides, which are acidic when dissolved in water. Such pursuit of anomalies was one reason why Cavendish was so slow to publish his findings, which he did some three years after the experiments were begun. But the fact is that Cavendish was in no hurry in any case. For him, publication was not the objective, and he seems blithely unconcerned about securing any claims to priority. He seems to have adopted the approach advocated by his colleague William Heberden, who said that the happiest writer wrote 'always with a view to publishing, though without ever doing so'.

So what was this experiment telling him? In retrospect it seems obvious: inflammable air and the 'active' constituent of common air – or hydrogen and oxygen, as Lavoisier was already calling these substances – unite to form water. But Cavendish was at that stage still in thrall to the phlogiston theory, and so things were by no means so clear to him. He ascertained that the lost one-fifth of common air could be identified as the 'dephlogisticated air' that Priestley had described: indeed, when he used this air instead of common air, it was all used up if ignited with twice its volume of inflammable air. But this inter-pretation meant that phlogiston had to appear somewhere in the balance. Cavendish concluded that dephlogisticated air was 'in reality nothing but dephlogisticated water, or water deprived of its phlogiston'. In other words, water was not being made from two constitutive parts but was appearing through the combination of phlogiston-poor water with the phlogiston contained in inflammable air. Alternatively, if the inflammable air were not phlogiston itself, then it was 'phlogisticated water', or 'water united to phlogiston'. The phlogiston effectively cancelled out:

$$[\text{water} - \text{phlogiston}] + [\text{water} + \text{phlogiston}] = \text{water}$$

How we are to understand Cavendish's conclusions has been a matter of great debate, because to some extent the issue of whether or not he made a genuine 'discovery' about the nature of water hinges on it. The truth is that there is nothing in what Cavendish wrote about his experiment that indicates unambiguously that he questioned the elemental status of water. That is to say, it remains unclear whether he decided that water somehow pre-existed in his airs and was simply being condensed in the explosion (which is pretty evidently what Priestley believed) or whether he had some inkling that water was being *created* from its constituents in a chemical process. Traditional historical accounts of Cavendish's experiment tend to imply that he made more or less the correct interpretation, even if he couched it in the archaic terms of phlogiston theory. But historian of science David Philip Miller has argued fairly persuasively that Cavendish's thoughts were closer to Priestley's. In any event, for an explicit and decisive statement of water's compound nature, we must look across the English Channel.

A new kind of chemistry

In Paris, Antoine Lavoisier was on the same path: familiar with Macquer's work, he too was looking more closely at what happened when the two airs were united. But he had a different hypothesis. In the mid-1770s he had concluded that Priestley's dephlogisticated air was in fact a substance in its own right: an element, which he proposed to call oxygen. The name means 'acid-former', for Lavoisier had the (misguided) notion that this element was the 'principle of acidity', the substance that creates all acids.

Cavendish knew of Lavoisier's oxygen, but he did not much care for it. He pointed out, quite correctly, that there was at least one acid – marine acid, now called hydrochloric acid – that did not appear to contain this putative element. (Lavoisier admitted in 1783 that there were some difficulties in that regard which he was still working on.) But while some of Cavendish's contemporaries, Priestley in particular, were trenchantly opposed to Lavoisier's theory because of an innate conservatism, Cavendish was more pragmatic – he argued simply that no one could at that stage know the truth of the matter. His objections were directed more at the way Lavoisier sought to impose the oxygen theory on chemical science by a relabelling exercise: in 1787 the French chemist proposed a new system of nomenclature in his magisterial *Traité elementaire de chimie*, the adoption of which would make it virtually impossible to practice chemistry without implicity endorsing oxygen. Imagine what would happen, Cavendish complained, if everyone who came up with a new theory concocted a new terminology to go along with it. In the end, chemistry would become a veritable Tower of Babel in which no one could understand anyone else. He derided the 'rage of name-making' and dismissed Lavoisier's *Traité* as a mere 'fashion'. Until there were experimental results that could settle such disputes, he said, it was better to stick with the tried-and-tested terminology, since new names inevitably prejudice the very terms within which theoretical questions can be framed.

That Cavendish's opposition was not motivated by mere traditionalism is clear from the fact that he gradually abandoned phlogiston and accepted Lavoisier's oxygen as the evidence stacked up in the French chemist's favour. Even in 1785 he was prepared to concede that phlogiston was a 'doubtful point', and by early 1787 the phlogistonist Richard Kirwan in England wrote to Louis Bernard Guyton de Morveau, a colleague of Lavoisier's, saying that 'Mr Cavendish has renounced phlogiston.' By the turn of the century, Cavendish was prepared even to use Lavoisier's terms: dephlogisticated air became oxygen, and inflammable air was hydrogen – the gas which, thanks to the researches of Warltire, Priestley and Cavendish as well as his own, Lavoisier saw fit to call the 'water former'.

But that is rather leaping ahead of the matter. In the late 1770s Lavoisier decided that, since his oxygen was the principle of acidity, its combination with hydrogen should produce an acid. In 1781–2 he looked for it in experiments along the same lines as Priestley and Warltire, but saw none. Working with Laplace, he combined oxygen and hydrogen in a glass vessel and found that their combined weight was more or less equal to that of the resulting water.

They were not the only French scientists to try it. When Joseph Priestley conducted further experiments of this kind in March 1783, the French scientist Edmond Charles Genet in London wrote a letter describing the work to the French Académie des Sciences, the equivalent of the Royal Society. Genet's letter was read to the academicians in early May. Lavoisier was there to hear it, and so was the mathematician Gaspard Monge from the military school of Mézières, who promptly repeated the experiment in June.

Lavoisier and Laplace did likewise – but by then they knew of Cavendish's results too, for Charles Blagden told them about his colleague's investigations in early June while on a trip to Paris. Lavoisier, as ambitious as Cavendish was diffident, quickly repeated the measurements on 24 June (forgoing, in haste, his usual quantitative precision) and presented them soon after to the Académie. He referred to the earlier work by both Monge and Cavendish, magisterially indicating that he 'proposed to confirm' Cavendish's observations 'in order to give it greater authority'.

Curiously, Lavoisier continued at this point to call oxygen 'dephlogisticated air' – but for him this was more or less just a conventional label, and did not oblige him to fit phlogiston into his explanations. That enabled him to see through to the proper conclusion with far more directness and insight than Cavendish. 'It is difficult to refuse to recognize', he said, 'that in this experiment, water is made artificially and from scratch.'

And, in a master stroke, he verified that this was so by showing how water might be split into its two constituents. Lavoisier felt that the right way to investigate the composition of matter was to come at it from both directions: synthesis, or making a substance from elemental components, and analysis, which meant separating the substance into those fundamental ingredients. He described how, in collaboration with the engineer Jean Baptiste Meusnier, he 'analysed' water by placing it together with iron filings in an environment free of air, held in an inverted bowl under a pool of mercury. The iron, he reported, was converted into rust, just as it is when it absorbs dephlogisticated air (that is, oxygen) from common air; and 'at the same time it released a quantity of inflammable air in proportion to the quantity of dephlogisticated air which had been absorbed by the iron.' 'Thus', he concluded, "water, in this experiment, is decomposed into two distinct substances, dephlogisticated air ... and inflammable air. Water is not a simple substance at all, not properly called an element, as had always been thought."

Cavendish's experiment was beautiful because of his attention to detail, a characteristic that redirected attention towards the formation of water and pointed clearly to the conclusion that Lavoisier subsequently drew. But Lavoisier's follow-up studies surely deserve a share of that beauty, because of the way he found the right interpretation and then went on to make it irrefutable.

Needless to say, not everyone saw it quite like that. The shroud of phlogiston that made Cavendish's explanation of his experiment somewhat ambiguous also helped to protect him from the kind of reactionary responses that Lavoisier's starker message attracted. An English chemist named William Ford Stevenson showed how reluctant some scientists were to abandon the ancient elemental status of water when he called Lavoisier's claims a kind of 'deception'. How on Earth could water, which puts out fires and was for that reason 'the most powerful antiphlogistic we possess', how could this substance truly be compounded from an air 'which surpasses all other substances in its inflammability'? Cavendish betrayed that he had not quite grasped the true implications of his results when he too expressed doubts about Lavoisier's conclusions. Priestley, a staunch believer in phlogiston, had no time for them. Blagden, meanwhile, was more angered (and with some justification) by Lavoisier's failure to give sufficient credit to what Cavendish had already achieved – although at that point Cavendish had still not submitted his report to the Royal Society.

Water wars

Even that was not the full extent of the controversy. No sooner had Cavendish's paper finally been read to the Royal Society in January 1784 than it awakened a new dispute. The Swiss scientist Jean André De Luc heard about the report and asked Cavendish for a copy, whereupon he wrote to his friend James Watt, suggesting that Cavendish was a plagiarist who had copied Watt's ideas 'word for word'. For Watt had repeated Warltire's experiment several years earlier while he was still a university technician at Edinburgh, working under Joseph Black. It was not so much the experiment itself that incited De Luc's charges, but Cavendish's interpretation in terms of 'dephlogisticated water', which seemed very much along the lines of what Watt had deduced: he had claimed that water was a compound of pure air and phlogiston.

At least, that is what some historical accounts indicate; but again, there is ambiguity about whether Watt truly identified water as a substance produced by the chemical reaction of two 'elements'. Drawing on Joseph Priestley's experiments in early 1783 on the spark ignition of dephlogisticated and inflammable air (which were themselves stimulated by Cavendish's still unpublished work), Watt suggested in April of that year that 'water is composed of dephlogisticated and

inflammable air, or phlogiston, deprived of part of their latent heat.'
Is this a statement that water is a compound substance? Latent heat is
the heat a gas releases when it condenses into a liquid – and so Watt's
conclusion seems to blend notions about both the combination of two
gases and the condensation of water. It's hard to know quite what to
make of it.

At that time, Watt expressed his ideas about water in letters to
Priestley, De Luc and Joseph Black. He had intended that they be
read out formally to the Royal Society, but then withdrew his formal
communication after learning that Priestley's further investigations
seemed to point to some inconsistencies with other ideas that Watt's
letter contained. Yet when De Luc saw Cavendish's report, he decided
that it had appropriated Watt's 'theory' without attribution.

Watt was annoyed, although unable to conclude for sure that
intellectual theft was involved. 'I by no means wish', he wrote to De Luc,

> to make any illiberal attack on Mr C. It is barely possible he may have
> heard nothing of my theory; but as the Frenchman said when he found a
> man in bed with his wife, 'I suspect something'.

All the same, Watt conceded that Cavendish's interpretations were
not identical to his own, and even admitted that 'his is more likely to
be [right], as he has made many more experiments, and, consequently
has more facts to argue upon'. There is a trace of envy at Cavendish's
riches and status in comparison to Watt's own humble origins (he was
the son of a Clydeside shipbuilder) when he tells De Luc that he 'could
despise the united power of the illustrious house of Cavendish'. Yet
Watt seems to have put aside his bitterness soon enough. He wrote a
paper that same year describing his own experiments and ideas on
water, in which he graciously noted that 'I believe that Mr Cavendish
was the first who discovered that the combination of dephlogisticated
and inflammable air produced moisture on the sides of the glass in
which they were fired.' The unworldly Cavendish probably never
knew about Watt's initial anger; in 1785 he recommended Watt for a
fellowship of the Royal Society.

This apparent conciliation did not prevent others from arguing over
who discovered that water was a compound. Cavendish, Watt, Lavoisier
and Monge have all been put forward as candidates. The debate raged
heatedly in the mid-nineteenth century, when it centred on Watt's rival
claim. His case was argued forcefully by François Arago, secretary of

the French Académie des Sciences, in his *Eloge de James Watt*, and Lord Brougham and Watt's son James Watt Jr added their voices to this appeal. In response, William Vernon Harcourt, in his address as president-elect to the British Association for the Advancement of Science in Birmingham in August 1839, vehemently asserted Cavendish's priority – a speech that left some members of the audience bristling, for Watt the engineer was a hero in the industrial Midlands of England.

David Philip Miller has shown that this 'water controversy' was fuelled by broader agendas. Watt Jr was no doubt motivated by filial concern for his father's reputation, but Arago and Watt's other supporters hoped that their protagonist's claim to this discovery in fundamental science would lend weight to their belief in an intimate link between pure and applied science. Harcourt's camp, meanwhile, consisted of an academic élite that was keen to promote the image of the 'gentleman of science' who sought knowledge for its own sake and remained aloof from the practical concerns of the engineer. The reclusive, high-born and disinterested Cavendish was its ideal exemplar. George Wilson's biography of Cavendish was a product of this controversy – a polemic that aimed to establishing its subject's priority and honourable conduct, it gave disproportionate attention to his experiments on water. But in other respects Wilson's researches left him with a rather poor impression of Cavendish's character, and his portrait set the template for the subsequent descriptions of a peculiar, asocial man, 'the personification and embodiment of a cold, unimpassioned intellectuality' as the editor of a collection of Cavendish's papers put it.

There is, however, a postscript to Cavendish's compulsive attention to detail that illustrates just how valuable to science pedantry can be. In 1783 he looked at the other component of common air, the 'phlogisticated air' that would not support combustion. This is, of course, nitrogen, which, as Cavendish showed, is converted into nitrous acid *via* reactions with oxygen. 'Acid in aerial form' was how Blagden summarized Cavendish's conclusions about phlogisticated air, and both he and Priestley felt that these studies represented a more important contribution than Cavendish's work on water – a reflection of the 'pneumatic' preoccupation of chemists at that time.

But while Cavendish was able to eliminate nearly all of the phlogisticated component of common air, he remarked that there always seemed to be a tiny bit of 'air' left, which appeared as a recalcitrant bubble in his experiments. This seemed to make up just $^1/_{120}$ part of common air, and Cavendish suspected that it was just the consequence of his experimental inadequacies. All the same, he wrote down his observations and Wilson mentioned them in the biography.

Some years later, an English chemistry student named William Ramsay bought a second-hand copy of Wilson's book and read about the mysterious bubble. That reference lodged in his remarkable mind, and he recalled it in the 1890s when he was a professor of chemistry at University College, London. Ramsay was at that time corresponding with the English physicist Lord Rayleigh, who suspected that nitrogen extracted from air might have a small impurity of some unreactive substance. They repeated Cavendish's experiments on nitrogen, and in 1894 they announced that they had discovered a new element, one that did not seem to react with any other. They named it after the Greek world for 'idle': argon. Within several years, Ramsay had found three other, similarly inert, gases and had unearthed an entirely new group of the periodic table of elements. That was the start of another story, and in Chapter 8 we shall hear its conclusion.

CHAPTER 3

New Light

The Curies' Radium and the Beauty of Patience

Paris, 1898–1902—In a cold and damp wooden shed at the School of Chemistry and Physics, the Polish scientist Marie Curie, occasionally assisted by her French husband Pierre, crushes and grinds and cooks literally tons of processed pitchblende, the dirty brown material left over from the mining of uranium. The Curies are convinced that this unpromising substance contains two new elements, which they name polonium and radium. After endless chemical extractions, Marie obtains solutions that glow with pale blue-green light: a sign that they contain the intensely radioactive element radium. The significance of the work is recognized straight away, as Marie and Pierre, still struggling to forge scientific careers in France, are awarded the Nobel prize in physics in 1903.

Marie Curie may well have felt ambivalent about becoming an icon for women's place in science. Like several trail-blazing women scientists, she seemed eager that her sex be seen as irrelevant. Sadly, it was not. Those scientists, such as Albert Einstein and Ernest Rutherford, who accepted Marie without question as an equal, stand out for their lack of prejudice; most of the scientific community at the end of the nineteenth century was reluctant to believe that a woman could contribute new, bold and original ideas to science. When Marie was grudgingly awarded prizes for her groundbreaking studies of radioactivity, as likely as not the news would be communicated via her husband. It was only by a hair's breadth that she was included in the decision of the 1903 Nobel committee. And when the Paris newspapers discovered that, several years after Pierre's death, Marie had had an affair with one of his former colleagues, they bit on the scandal with relish, when similar behaviour from an eminent male scientist would probably have been deemed too trivial to mention.

It is hardly surprising, then, that Marie took great pride in her work, carefully emphasizing her own contributions and hastening to publish them in the face of Pierre's habitual indifference to public acclaim. Marie knew that, to make her mark, she would have to achieve twice as much as her male colleagues. And she did – which is why she became the only scientist to win two Nobel prizes in science.

Her life has been so often romanticised – that process began as soon as the news came from Stockholm in 1903 – that Marie Curie herself has tended to disappear behind the stereotype of the tragic heroine. Yet it is true that her life was marred by several tragedies and by considerable adversity, and it is not surprising that in the end this left her hardened, appearing aloof and cold to those around her. Her determination and dedication to her work could translate as a certain prickliness and unfriendliness towards her colleagues. If she expected others to ignore the fact that she was a woman, likewise she herself had no concern about protecting brittle male egos.

In as much as she discovered new elements, Marie Curie did nothing that others had not done previously. But the elements that she unearthed in her long and arduous experiment were like nothing anyone had seen before.

New physics

The elements that debuted in the periodic table in the late nineteenth and early twentieth century hint at the unattractive nationalism of that age: they bear names like gallium, germanium, scandium, francium. We can hardly begrudge Marie Curie her polonium, however, since her own sense of national pride was born out of Russian oppression. Poland was then part of the Russian empire, and after a rebellion in 1863 (four years before Marie's birth) the country suffered from an intensive programme of 'Russification', during which the tsar forbade the use of the Polish language in official circles. The struggle of the Polish intelligentsia against the Russian authorities was a dangerous business in which some lost their lives.

Figure 3 *Marie Curie (1867–1934), the discoverer of radium and polonium*
(© CORBIS)

When Maria Sklodowska (Figure 3) came to Paris to study science
and mathematics at the Sorbonne, it must have seemed a land of
opportunity – this was the Paris of Debussy, Mallarmé, Zola, Vuillard
and Toulouse-Lautrec. Yet women risked their reputation simply by
venturing out into the city alone, and Maria was more or less confined
to her garret lodgings in the Quartier Latin. She graduated in 1894 and
began working for her doctorate under the physicist Gabriel
Lippmann, who later won the Nobel prize for his innovations in
colour photography.

That year she met the 35-year-old Pierre Curie, who taught at the School of Chemistry and Physics and studied the symmetry properties of crystals. Pierre did not possess the right credentials to become part of the French scientific élite – he had studied at neither of the prestigious schools, the École Normale or the École Polytechnique – but nonetheless he had made a significant discovery in his early career. With his brother Jacques at the Sorbonne in 1880, he had found the phenomenon of piezoelectricity. When the mineral quartz is squeezed, the Curie brothers discovered, an electric field is generated within it. They used this effect to make the quartz balance, which was capable of measuring extremely small quantities of electrical charge. Pierre's colleague Paul Langevin used piezoelectricity to develop sonar technology during the First World War.

Shy and rather awkward in public, Pierre had never married. His work was almost an obsession and he did not seem interested in acquiring a token wife. 'Women of genius are rare', he lamented in his diary in 1881. But he quickly recognized that Maria Sklodowska was just that kind of rarity, and in the summer of 1894, when she had returned temporarily to Poland, he wrote to her: 'It would be a beautiful thing, a thing I dare not hope, if we could spend our life near each other hypnotized by our dreams: your patriotic dream, our humanitarian dream and our scientific dream'.

His courtship was a little unorthodox – he dedicated to Maria his paper 'Symmetry in physical phenomena'. But it seemed to work: they were married in 1895, the same year in which Pierre (never one to rush his research) completed his doctoral thesis on magnetism.* The marriage delayed Marie (as she now called herself) from starting on her own doctorate, for she soon had a daughter, Irène, who was later to become a Nobel laureate too. This hiatus turned out to be doubly fertile, for Marie's eventual research topic was a phenomenon discovered only in March 1896.

The fin-du-siècle produced something of a public fad for the latest science and technology. Gustave Eiffel's steel tower rose above the Paris skyline in 1889 as a monument to technological modernism, and was quickly embraced as such by Parisian artists like Raoul Dufy and Robert Delaunay. Emile Zola claimed to be writing novels with a scientific spirit, and his book *Lourdes* (1894), which Pierre gave to Marie, was a staunch defence of science against religious mysticism.

---

* The temperature at which magnets lose their magnetism is now called the Curie point.

When Wilhelm Conrad Roentgen discovered X-rays in 1895 and found that they could 'look inside' matter by imprinting a person's skeleton on a photographic plate, the public's imagination was quickly captured – the Paris carnival parade of 1897 even had an 'X-ray float'.

Roentgen made his discovery while investigating so-called cathode rays, which were emitted from negatively charged metal electrodes. These mysterious rays were typically produced in a sealed glass tube containing gases at very low pressure: the cathode ray tube. In 1895 the French physicist Jean Perrin, later a firm friend of the Curies, showed that cathode rays deposited electrical charge when they struck a surface. J. J. Thomson at Cambridge showed two years later that cathode rays were deflected by electric fields, and he concluded that they were in fact streams of electrically charged particles, which he called electrons.

Roentgen was studying cathode rays in 1895 when he noticed that some rays seemed to escape from the glass tube, causing a nearby phosphorescent screen to glow. This effect had already been noted previously by the German physicist Philipp Lenard, but Roentgen investigated it more closely. He found that the rays were capable of penetrating black cardboard placed around the tube. And when he placed his hand in front of the glowing screen, he saw in shadow a crude outline of his bones. In December of 1895 he showed that the rays would trigger the darkening of photographic emulsion, and in that way he took a photograph of the skeleton of his wife's left hand.

These were evidently not cathode rays. It was already known that cathode rays were deflected by magnets, but a magnetic field had no effect on these new, penetrating rays. Roentgen called them X-rays, and scientists soon deduced that they were a form of electromagnetic radiation: like light, but with a shorter wavelength. The French scientist Henri Poincaré described Roentgen's discovery to the Académie des Sciences in January 1896, and among those who heard his report was Henri Becquerel. Becquerel's father had made extensive investigations of phosphorescence – the dim glow emitted by some materials after they have been illuminated and then plunged into darkness – and he wondered 'whether ... all phosphorescent bodies would not emit similar rays'. This was actually a rather strange hypothesis, for the phosphors on Roentgen's screens were clearly *receiving* X-rays, not *emitting* them. All the same, Becquerel went looking for X-rays from phosphorescent materials.

That February he wrapped photographic plates in black paper and then placed phosphorescent substances on top and exposed them to the sun to stimulate their emission. But most of these materials generated no sign of X-rays – the plates stayed blank. Uranium salts, however, would imprint the developed plates with their own 'shadow'. At first, Becquerel assumed that sunlight was needed to cause this effect, since after all that was what induced phosphorescence. He set up one experiment in which a copper foil cross was placed between the uranium salt and the plate, expecting that the foil would shield the photographic emulsion from the rays apparently emanating from uranium. A shadow of the cross should then be imprinted on the developed plate. But February is seldom a sunny month in northern Europe, and on the day that Becquerel set out to perform this experiment the sky was overcast. So he put the apparatus in a cupboard for later use. But the weather remained gloomy, and after several days Becquerel gave up. Again we have cause to be thankful for the fluid logic of Becquerel's mind, for rather than just writing the experiment off and casting the photographic plate aside, he went ahead and developed it anyway. The uranium had received a little of the winter sun's diffuse rays, after all, so there might at least be some kind of feeble image in the emulsion.

To his amazement, he found that 'on the contrary, the silhouettes [of the copper mask] appeared with great intensity'. Thus, sunlight wasn't needed to stimulate the 'uranic rays'. Still in thrall to the idea of phosphorescence, Becquerel dubbed this 'invisible phosphorescence' or hyperphosphorescence.

At first his discovery made little impact. These 'uranic rays' were too weak to take good skeletal photographs, and most scientists remained more interested in X-rays. The Curies, however, recognized that Becquerel's result was pointing to something quite unprecedented, and in early 1898 Marie decided to make this the topic of her doctorate. 'The subject seemed to us very attractive', she later wrote, 'and all the more so because the question was entirely new and nothing yet had been written upon it'.

Return to the source

It was very much a joint project, which the Curies began in an empty store room of the School of Chemistry and Physics. They first found a method of quantifying the 'activity' of the uranic emissions by measuring their charging effect on a metal electrode. Becquerel had commented that the rays made air electrically conducting – as we'd now say, they ionize the air, knocking electrons out of the atoms and leaving them electrically charged. Pierre's piezoelectric quartz balance now came into its own for measuring the amount of charge deposited on a metal plate due to a sample of uranium salt placed below it.

At first the Curies used relatively 'pure' materials: uranium salts given to them by the French chemist Henri Moissan. But in February 1898 Marie tested raw pitchblende – uranium ore, which was mined in the town of Joachimsthal in Saxony, where silver mining had been conducted since the Middle Ages. Remarkably, crude pitchblende turned out to be even more active than purified uranium. Likewise, whereas salts of the rare element thorium were also found to emit 'uranic rays', the raw mineral form of thorium (aeschynite) was more active than pure thorium compounds.

The Curies had a crucial insight: they hypothesized that the greater ionizing power of pitchblende was caused by an unknown element, more 'active' than uranium itself, which was present as an impurity in the mineral. To verify this, they compared another natural uranium mineral, chalcite, with 'artificial' chalcite synthesized chemically from uranium and copper phosphate. Superficially, the two materials should be identical; but the synthetic chalcite had only uranium-like activity, whereas natural chalcite was more active. So there was something else in this mineral too: some ingredient with a 'uranic' potency exceeding that of uranium. What they needed to do was to isolate it.

The Curies reported their findings and hypothesis to the Académie on 12 April. In effect, this report suggested that radioactivity could be used as a diagnostic signal to search for new elements: invisible to chemical analysis, the hypothetical new source of uranic rays betrayed its presence by its emission. 'I had a passionate desire to verify this new hypothesis as rapidly as possible', Marie wrote.

'Passionate' is not a word commonly associated with Marie Curie. She had been brought up to observe the genteel, reserved manners expected of a lady of that era. Even Einstein, who was fond of Marie, confessed that he found her 'poor when it comes to the art of joy and pain'. There can be no doubt, judging from her own words, that she was devoted to her husband and her children, and her pain at the tragedies in her later life is clear and deeply felt. But she would, if she could, keep her passions for other people very private. The comment of *Le Journal* in 1911 on her affair with Paul Langevin was an example of pure tabloid lasciviousness – 'The fire of radium had lit a flame in the heart of a scientist' – and was met by her justifiably icy response in *Le Temps*: 'I consider all intrusions of the press and of the public into my private life as abominable'. The only passion that Marie Curie permitted herself to reveal publicly was that for her work, and like Pierre's, it bordered on obsession.

If there was a new element lurking in pitchblende, it would have to be separated by chemical ingenuity, and the Curies enlisted the help of a chemist named Gustave Bémont at the School of Chemistry and Physics. Two dissolved elements may be parted if one of them forms an insoluble compound while the other does not: the one can be precipitated and collected by filtering, while the other remains dissolved in solution. In such a procedure, an element present in only very small amounts can sometimes be separated by precipitating it along with some other element with which it shares chemical properties in common: the trace element gets entrained with the 'carrier'. The Curies found that in fact pitchblende seemed to contain *two* new 'active' elements. One of them was chemically similar to barium, precipitating when chloride was added to a solution of the mixture to produce insoluble barium chloride. The other element seemed instead to 'follow' the element bismuth.

These separations involved laborious, repetitive procedures in which chemical products were crystallized from solution, washed and redissolved and then recrystallized. It was tedious, mind-numbing work. But the Curies tracked the progress of their labours by using the ionization apparatus to measure changes in the 'activity' of their samples. Since the new elements were more active than uranium, products in which they were enriched relative to pitchblende showed greater activity. By the summer of 1898 the Curies had increased the activity of their extract by a factor of around 300. They hoped that the new elements might be revealed in these enriched samples by the technique of spectroscopy. Elements irradiated by light re-emit some of the light in the form of distinct 'spectral lines' at specific wavelengths – this was how the element helium was discovered in 1868, when astronomers found previously unknown spectral lines in sunlight (the sun is rich in helium). But when the Curies gave their enriched samples to the French scientist Eugene Demarçay for analysis, he was unable to find any new spectral lines. There was still more purification to be done before the highly active elements would show themselves that way.

This did not prevent the Curies from presenting their findings to the Institut de France in July, in a paper read by Becquerel. That month, they had chosen a name for one of the new elements they were sure the samples contained. 'We thus believe', they said, 'that the substance we have extracted from pitchblende contains a metal never before known, akin to bismuth in its analytical properties. If the existence of this metal is confirmed, we propose to call it *polonium* after the name of the country of origin of one of us'. The paper's title introduced another new word: 'On a new radio-active substance contained in pitchblende.'

But it was the other new element that the Curies decided to pursue: not polonium, but the residue that followed barium. They finally prepared a sample with 900 times more radioactivity than pitchblende, and Demarçay saw at last a new spectral line. Here was their evidence. And to their delight, they found that solutions enriched in this element had an unexpected attraction. 'We had an especial joy', Marie wrote, 'in observing that our products containing concentrated radium were all spontaneously luminous. My husband who had hoped to see them show beautiful colourations had to agree that this other unhoped-for characteristic gave him even greater satisfaction'. The luminous property of the new element suggested to them the name that Pierre recorded in his lab notebook around 20 December 1898: radium. The day after Christmas, the Curies and Bémont presented a paper titled almost identically to the earlier one: 'On a new strongly radio-active substance contained in pitchblende.'

A new spectral line was proof enough for physicists. But chemists expect their materials to be tangible things. If you think you have a new element, they might have said in those days, show it to me. Place some in my hand. Marie realised that the only way she would have a really secure claim would be to isolate pure radium. So that is what she set out to do.

Radiant rubbish

It was a reductive process: they started with a mass of pitchblende, and boiled and crystallized and recrystallized it, and with each enrichment step the quantity of material diminished. If they wanted to get enough of the new elements to ensure their positive identification, they'd need to process vast quantities of pitchblende. The raw material was valuable stuff: it was the source of uranium, which was used to make a commercial orange glaze for ceramics. But the Curies didn't care about uranium – what they wanted was the tiny amount of radium that was presumably left behind once the uranium was extracted. To their delight, they verified that the pitchblende residue from uranium mining was indeed more radioactive than pitchblende itself: removal of uranium had already concentrated the precious radium. To the Joachimsthal miners, this residue was worthless: the brown powder was simply dumped in the pine forests. Even so, shifting huge quantities of the stuff from Saxony to Paris could not happen for free, and it was only thanks to donations from Baron Edmond de Rothschild that the Curies were able to procure ten tons of it. It arrived in sacks of 'brown dust mixed with pine needles', a most unpromising ingredient. The words of the alchemists in search of the Philosopher's Stone might have given the Curies encouragement, had they but known of them: 'The Sages say that their substance is found on the dung hill'.

Processing this quantity of material required plenty of space, and so the Curies needed a new laboratory. What they were given, at the School of Chemistry and Physics, had the virtue of capaciousness but little else. Marie later described it as

> a wooden shed with a bituminous floor and a glass roof which did not keep the rain out, and without any interior arrangements. The only objects it contained were some worn pine tables, a cast-iron stove, which worked badly, and the blackboard which Pierre Curie loved to use. There were no hoods to carry away the poisonous gases thrown off by our chemical treatments, so that it was necessary to carry them on outside in the court, but when the weather was unfavourable we went on with them inside, leaving the windows open.

Irène Curie later described the task that Marie set herself:

> My mother ... had no fear of throwing herself, without personnel, without money, without supplies, with a warehouse for a laboratory, into the daunting task of treating kilos of pitchblende in order to concentrate and isolate radium.

As news of the herculean efforts of the Curies spread, they began to receive eminent visitors in their Paris shed. Lord Kelvin, the most famous British physicist of the Victorian era, was a particularly supportive patron, and one day the German scientist and future Nobel laureate Wilhelm Ostwald stopped by when the Curies were not there. He was astonished to see their workplace: 'It was a cross between a stable and a potato-cellar, and, if I had not seen the worktable with the chemical apparatus, I would have thought it a practical joke'.

The bulk of the work on isolating radium was conducted by Marie alone; her husband focused on trying to understand the phenomenon of radioactivity. This division of the manual and routine from the abstract and novel did not necessarily reflect their respective strengths, for Marie was as adept at mathematics as Pierre; it is rather tempting to see it now as a reflection of the prejudice that said a woman's place was in the kitchen. But Pierre was dogged by aches and pains that he attributed to incipient rheumatism, while Marie proved quite capable of (even enthusiastic about) the exhausting physical labour required to reduce tons of pitchblende residue to 'great vessels full of precipitates and of liquids'. 'Sometimes', she said, 'I had to spend a whole day mixing a boiling mass with a heavy iron rod nearly as large as myself. I would be broken with fatigue at the day's end'. All the same, she felt that she was working her way ever closer to her goal, and she took delight in her progress:

> We were very happy in spite of the difficult conditions under which we worked. We passed our days at the laboratory, often eating a simple student's lunch there. A great tranquillity reigned in our poor shabby hangar; occasionally, while observing an operation, we would walk up and down talking of our work, present and future. When we were cold, a cup of hot tea, drunk beside the stove, cheered us. We lived in a preoccupation as complete as that of a dream.

There was a rough-hewn romance to it, without doubt. But could this mechanical extraction of radiant matter from great, steaming vats of noxious raw material, reminiscent of the alchemist Hennig Brandt's isolation of phosphorus from urine in the seventeenth century (page 97), really be deemed beautiful? Well, we need only picture the Curies standing in their shed at dusk after another day's back-breaking effort to appreciate why this years-long experiment is indeed one of the most ravishing in the history of science:

> One of our joys was to go into our workroom at night, we then perceived on all sides the feebly luminous silhouettes of the bottles or capsules containing our products. It was really a lovely sight and always new to us. The glowing tubes looked like faint fairy lights ... these gleamings, which seemed suspended in darkness, stirred us with ever new emotion and enchantment.

The energy inside the atom

In July 1902, Marie Curie reported the fruits of her efforts: one-tenth of a gram of a pure radium compound. 'It had taken me almost four years', she wrote, 'to produce the kind of evidence which chemical science demands, that radium is truly a new element.' Radium turned out to exceed the radioactivity of uranium by a far greater margin than anyone had expected: a factor of around a million. Marie was finally in a position to measure its density, and thus to deduce its atomic weight. This quantity established the imprimatur of a true element: a place in Dmitri Mendeleyev's periodic table. She reported in *Comptes rendus* that the atomic weight of radium was 225; the accepted value today is 226. 'According to its atomic weight', she concluded, 'it should be placed in the Mendeleev table after barium in the column of alkaline earth metals' – a conclusion that she would have guessed in any case in view of radium's tendency to behave like barium (elements in the same column typically share the same chemical properties; see page 74). The speck of material that Marie extracted was a salt of radium, a combination of elements. It took another nine years to collect enough of one such compound to extract from it the pure, metallic element by electrolysis, something that Marie accomplished with André Debierne.

'It is not an exaggeration' Jean Perrin claimed in 1926, 'to say that [the isolation of radium] is the cornerstone on which the entire edifice of radioactivity rests.' That edifice had, however, been rising steadily over the years that Marie worked away in her shed in Paris; and the Curies played a central role in its construction. When they presented a paper called 'The new radioactive substances' at the International Congress of Physics in Paris in 1900 before such luminaries as Kelvin, Hendrik Lorentz, Jacobus van't Hoff and Svante Arrhenius, it excited intense interest. The key question was: where did the energy of radioactive substances come from? Clearly, the 'rays' emitted by these materials carried energy: the glow of radium was visible testament that the atoms of this element were radiating energy like tiny, feeble suns. Marie was baffled by the question. She noted that the emission of uranic rays 'does not vary noticeably with time', and meanwhile 'the uranium shows no appreciable change of state, no visible chemical transformation, it remains, in appearance at least, the same as ever, the source of the energy it discharges remains undetectable'. 'That' she confessed, 'is the most troubling aspect of the phenomenon'. Indeed, it was so perplexing that she was even moved tentatively to suggest a virtual heresy: might radioactivity violate the first law of thermodynamics, the stipulation that energy can be neither created nor destroyed?

Another possibility was that somehow radioactive substances were soaking up and re-radiating energy conveyed to them from outside, by rays that permeated all space but could be absorbed only by heavy elements such as uranium and thorium. It sounds like (and in truth it was) grasping at straws, but the idea was taken seriously enough for two Germans, Julius Elster and Hans Geitel, to make measurements on radioactive samples underneath 300 metres of rock in the Harz mountains of Saxony, and at the foot of an 850-foot mineshaft. They reasoned that the rock would shield the substances from any hypo-thetical radioactivity-inducing rays. These were the first experiments conducted underground to exclude penetrating environmental effects, pre-empting today's searches for neutrinos and other exotic subatomic particles in deeply buried laboratories. But the two researchers saw no change in the activity of their radioactive samples, which led them to believe that the energy must be emanating 'from the atom itself'. Radioactivity was, in other words, a form of atomic energy.

Marie Curie had concluded almost as much already: 'radioactivity', she said, 'seems to be an atomic property, persistent in all the physical and chemical states of the material'. To understand radioactivity, one needed to look into the atom itself.

Marie submitted her doctoral thesis on 25 June 1903 – for we must recall, perhaps with some astonishment, that the discovery of two new elements and the four-year labour that left her the world expert on radioactivity was all in pursuit of a doctorate. She was awarded the qualification with a distinction *très honorable*. And at the celebration that evening in Paris her guests included the man who dissected the atom and revealed its hidden source of energy: the New Zealander Ernest Rutherford.

While at McGill University in Canada, Rutherford discovered in 1899 that there were two types of rays emitted from radioactive materials: alpha rays, which were stopped dead by thin aluminium foil, and beta rays, which were able to penetrate such a barrier. The Curies and Becquerel deduced that Rutherford's beta rays were identical to the cathode rays that Thomson had identified as charged particles: electrons. Also in 1899, Rutherford, puzzled by the need to enclose thorium in a metal box in order to make consistent measurements of its radioactivity, discovered that it gave off a radioactive gas or 'emanation'. He teamed up with British chemist Frederick Soddy to figure out what the 'thorium emanation' (which they called thoron) really was. Theirs was perhaps the most fertile of the several associations of a physicist and a chemist that characterized the early days of radiochemistry and atomic physics.

In late 1900, the two researchers found that 'thoron' was nothing other than the inert gas argon, which William Ramsay and Lord Rayleigh had discovered six years earlier (see page 36). They realized the implication with something akin to horror. 'The element thorium was slowly and spontaneously transmuting itself into argon gas!' Soddy later wrote. At the time, he was shocked. 'Rutherford', he reportedly stammered to his colleague in the lab, 'this is transmutation: the thorium is disintegrating.' 'For Mike's sake Soddy', Rutherford thundered back, 'don't call it transmutation. They'll have our heads off as alchemists.'

But transmutation was truly what it was: Rutherford and Soddy deduced that uranium, thorium and radium changed spontaneously into new chemical substances by emitting alpha rays. (In the next chapter we will see what Rutherford concluded about these rays.) This process released the mysterious energy of radioactivity, and in 1903 Soddy made an awesome estimate of how much of this energy was available:

> It may therefore be stated that the total energy of radiation during the disintegration of one gram of radium cannot be less than 10^8 gram-calories, and may be between 10^9 and 10^{10} gram-calories .... . The energy of radioactive change must therefore be at least twenty-thousand times, and may be a million times, as great as the energy of any molecular change.

A year later, in a talk on the uses of radium, Soddy warned that 'The man who put his hand on the lever by which a parsimonious nature regulates so jealously the output of this store of energy would possess a weapon by which he could destroy the earth if he chose'. But probably not even Soddy guessed that it would take just 40 more years to build that weapon, and to use it.

Pierre Curie came to the same conclusion. In that same year of 1904, when he delivered his Nobel lecture in Stockholm (having been unable to travel to Sweden for the prize ceremony the previous year), he said 'It is possible to conceive that in criminal hands radium might prove very dangerous, and the question therefore arises whether it be to the advantage of humanity to know the secrets of nature, whether we be sufficiently mature to profit by them, or whether that knowledge may not prove harmful'. We must each make up our own minds about the answer to Pierre Curie's question.

Fame and fate

The Nobel prize changed everything for the Curies, and not necessarily for the better. It alleviated their financial problems – for despite their growing international acclaim, they continued to suffer neglect from the conservative French scientific establishment, and grant money was scarce. (In 1900 Pierre finally got a post at the Sorbonne, thanks to the support of Henri Poincaré, but two years later his candidacy for election to the Académie des Sciences was turned down.) Indeed, the French press were outraged when the Nobel award highlighted how the Curies had been cold-shouldered: 'During seven years', said *La Grande Revue*, 'no one in our country thought to reward these admirable scientists who have made this new conquest .... It took the generosity of a foreigner.'

But such adulation carried with it an obligation to endure endless, and often mindless, interviews, and the reclusive Curies were utterly dismayed to find that serious work became impossible in the face of visits from journalists, invitations to public events, and streams of letters from beggars, cranks and social climbers. The award, Marie later admitted, 'had all the effects of disaster.' 'One would like to dig into the ground somewhere to find a little peace', she wrote at the time.

Radioactivity in general, and radium in particular, were now the new X-rays. Everyone was fascinated by the Curies' glowing substance; Pierre complained to his friend Georges Gouy about the 'sudden infatuation for radium'. Chemicals companies recognized the commercial value of the permanently luminous material – it was soon to feature on watches and instrument panels everywhere – and a Parisian firm called the Central Society of Chemical Products approached Marie as she strove to isolate the element, asking her about her methods and offering to scale up the separation process for industrial production. By 1900 there were two German companies making (impure) radium compounds. The casinos of New York sold their customers luminous chips for playing 'radium roulette', during which they were served glowing cocktails laced with radium. In San Francisco, showgirls caused a sensation by dancing in the dark in radium-painted costumes.

It wasn't just about the radium glow, however. The German scientist Friedrich Walkhoff had shown that X-rays could be used to treat cancerous tumours, and Pierre Curie suggested that radium might have a similar application. The notion began to circulate that radium had marvellous medical properties. America was (as ever it has been) a playground for medical quacks: the 'radium craze' began as soon as the Curies, along with Henri Becquerel, won their Nobel, and for three decades it spawned cures such as Raithor, a solution of radium salts that was advertised as a panacea, capable even of dispelling mental illness. The Pittsburgh businessman Eben Beyers drank a bottle of Raithor every day until, after four years, he died of cancer of the jaw. In 1914 the medical committee of the British Science Guild warned that 'there is a danger that the claims which have been advanced for radium as a curative agent may lead to frauds on the credulous section of the public, which may be imposed upon by the sale of substances or waters in which radium does not exist, or may be harmfully treated by persons with no medical qualifications'.

The Curies had reason enough to know that fraud was not the real problem. As they carried out their seminal work, they found themselves chronically afflicted by lethargy and constantly prone to minor ailments. Marie lost ten pounds during this time, prompting the Curies' friend George Sagnac to scold Pierre about how little they both ate. In 1903 Pierre wrote to the Scottish physicist James Dewar in London that 'Madame Curie is always tired without being exactly ill.' That summer, her second pregnancy ended with the premature birth of a baby who died soon after delivery. And in November she was diagnosed as anaemic. The Curies did not attend the Nobel ceremony (Becquerel accepted their prizes along with his own) because they were too ill.

They knew that radioactive materials could be harmful, but did not connect this with their long-term illnesses. They thought that the ill effects of radioactivity were limited to skin burns, as Pierre and Becquerel reported from their own personal experience in 1901. Pierre once strapped some radium-enriched barium to his arm for ten days, and observed that the skin grew red and then developed into a wound which was evident even after seven weeks, 'indicating deeper injury'. These effects could make it difficult and painful to handle radioactive solutions; the Curies recorded that 'the ends of fingers which had held tubes or capsules containing very active products became hard and sometimes very painful; for one of us the inflammation of the ends of the fingers lasted 15 days and ended with the shedding of skin, but the pain has not disappeared completely after two months.' Moreover, as was discovered in 1899, radioactive materials could induce radio-activity in initially 'inactive' substances – before long, the Curies' entire laboratory was measurably radioactive, and their notebooks are still too 'hot' for safe handling.

By 1906, Pierre was very ill – permanently tired, and also depressed. He hadn't published a paper for two years, and his joints ached with what he took to be rheumatism. One can only wonder what his fate would have been, had it not caught up with him prematurely; for on 19 April he was struck by a horse-drawn cart in the rue Dauphine and knocked under the iron-shod wheels, where his skull was shattered. He died instantly – and so too did something in Marie. She was appointed to her husband's former position at the Sorbonne (becoming its first female professor), where she lectured in a way that suggested she was no longer interested in life.

Yet her humanitarian concerns remained. During the First World War she operated a mobile X-ray unit, and subsequently she researched the uses of radium in cancer therapy. She could have had no doubt, however, of its Janus nature – she continued to endure radiation-burnt hands throughout her life, and on 4 July 1934 she died of leukaemia, probably brought on by the object of her scientific passion. She was, said Albert Einstein, 'of all celebrated beings, the only one whom fame has not corrupted'.

The Curies' 1903 Nobel prize was awarded in physics, 'in recognition of the extraordinary services they have rendered by their joint researches on the radiation phenomena discovered by Professor Henri Becquerel'. That award assured Marie's fame, but it is surely the 1911 prize in chemistry, given to Marie alone, that stands as testimony to her beautiful, tenacious and fatal experiment: 'in recognition of her services to the advancement of chemistry by the discovery of the elements radium and polonium, by the isolation of radium and the study of the nature and compounds of this remarkable element'.

CHAPTER 4

Radiation Explained

Rutherford's Alpha Particles and the Beauty of Elegance

Manchester, 1908—Ernest Rutherford satisfies his long obsession with the alpha particles emitted by radioactive materials – particles that he was the first to identify and to name – by demonstrating conclusively that they consist of the nuclei of helium atoms. This discovery helps to make sense of the unsettling fact that radioactive elements appear to undergo transmutation: to become converted, as they emit radiation, into entirely new chemical elements.

How could anyone fail to like Ernest Rutherford? He was the archetypal bluff New Zealander (Figure 4), a genial extrovert, tall, good-looking, boisterous and plain-spoken, always ready with a joke or a tuneless burst of song, a charming, kind and gallant man who also happened to possess a clear-sighted genius. It is said that he 'never made an enemy or lost a friend', although Paul Langevin, who lodged as a student next door to Rutherford in Cambridge, wondered whether one could really speak in terms of friendship with a man who was 'a force of nature'.

At the English universities his manner, learnt in the backwoods of the antipodes, was considered a little unrefined, a little lacking in grace. Rutherford was no stranger to disagreements, even with those people he befriended and respected, such as his mentor J. J. Thomson and his collaborator Frederick Soddy. He would sometimes suffer black moods or experience sudden rages in which he might dash a piece of equipment to the ground. He had no time for fools or bunglers. All of this, however, is part of what marks him out as one of the most human,

Figure 4 *New Zealand physicist Ernest Rutherford, one of the greatest experimental scientists of all time*
(Reproduced Courtesy of the Library and Information Centre, Royal Society of Chemistry)

most vibrant scientists of the modern age. Indeed, his friend the chemist Chaim Weizmann considered that Rutherford 'suggested anything but the scientist. He talked readily and vigorously on every subject under the sun, often without knowing anything about it.'

Rutherford delighted in experiment, all the more so when it called on his manual skills. The elegance of his practical work seems to have derived from a rare conjunction of clarity of conception and pleasure in finding a way to physically realise it. There appeared to be some kind of fertile feedback between Rutherford's hands and his mind, so that his ability to visualize and to fabricate a piece of apparatus helped the experiment to take shape in his head. He loved his gadgets, and was uncommonly hard on students or collaborators whose clumsiness led them to commit 'crimes against apparatus'. 'Rutherford was an artist', one of his students later said. 'All his experiments had style.' It is tempting to go further and claim that Rutherford was the greatest experimental scientist of the twentieth century.

All the same, he marks the end of an era. Not only did his work usher in the modern science of nuclear and subatomic physics, but it led inevitably to an age in which fundamental research was conducted by huge international teams using machines of awesome scale and power built on multi-million-dollar budgets. Rutherford would never have been comfortable among the gigantic particle accelerators and colliders needed today to continue the exploration into the Lilliputian guts of matter. At heart he was a Victorian, content to build his equipment with sealing wax and string, and at the end of his career this made him seem almost a hidebound curmudgeon, deploring the commercialization of science and turning down large industrial grants in the belief that the best science was done on a shoestring.

Rutherford turned the Curies' 'radiochemistry' into atomic physics. In 1909 he discovered that the atom was mostly empty space, with almost all of its mass concentrated in a tiny, dense nucleus. And by subsequently 'splitting the atom', as the newspapers called it in 1919, Rutherford took the study of matter beyond the regime that chemistry could accommodate; for the atom is chemistry's most fundamental unit, and once it is broken the vocabulary of the chemist becomes redundant – there are new rules, new divisions of matter, a whole new world to navigate.

But this chapter is not about those great experiments. It is about one that came before, one that arguably belongs to the legacy of chemistry – and which, for elegance of conception and execution, Rutherford never surpassed.

Chasing the alpha

It began for Rutherford in the same way as it did for the Curies: with Wilhelm Roentgen's X-rays. Rutherford came to Cambridge in 1895, the year that Roentgen made his discovery, and, like so many other scientists of that time, he was intrigued by it. Rutherford worked under J. J. Thomson, whose studies of cathode rays were then leading him to the discovery of the electron – the first *subatomic* particle, a hint that atoms were reducible. At first Rutherford studied how X-rays could ionize gases – that is to say, how they knock electrons off gas atoms. The researchers in Cambridge's Cavendish laboratory did not at that point have such a clear notion of what was involved – all they really knew was that X-rays improved the electrical conductivity of gases, enabling an electrically charged body to shed its charge by releasing a spark.

Radioactivity caused ionization too, and in 1897 Rutherford switched his attention to the effects of the mysterious radiation that came from uranium compounds. Like Pierre Curie, he used an electrical discharge apparatus to measure radioactivity: the more active a sample, the more readily it induced a discharge. Rutherford found that if he covered a uranium sample with aluminium foil, its activity declined. If he added another layer of foil, and then another, the radiation continued to decrease in intensity. But beyond that, more layers of foil made essentially no difference: the remaining level of radioactivity stayed the same. Rutherford concluded that the foil was absorbing one type of radiation while being quite unable to influence a second type. 'These experiments', he reported in 1899,

> show that the uranium radiation is complex and that there are present at least two distinct types of radiation – one that is very readily absorbed, which will be termed for convenience the alpha-radiation, and the other of more penetrative character which will be termed the beta-radiation.

That was a revelation. Radioactivity was already remarkable enough, but now it seemed as though there was even more to explain. There is some evidence that Rutherford detected the third type of radiation too, although he did not clearly identify it as such. This is gamma radiation, which is now known to be far more penetrating than, and different in kind from, both alpha and beta rays. The latter two are in fact particles, whereas gamma rays are electromagnetic waves, like X-rays.

'Alpha-radiation' was a more effective ionizer than 'beta-radiation', which made it easier to study quantitatively. This seems to have been the simple reason why Rutherford elected to focus his attention on it at the expense of 'beta rays' (which were shortly shown to be identical to cathode rays, J. J. Thomson's electrons). But alpha rays became Rutherford's consuming passion, to the extent that he considered them 'his' and often referred to them as such.

Like others working on radioactivity, Rutherford was baffled, even disturbed, by all this energy streaming spontaneously out of a natural mineral like pitchblende. In 1900, two years after he moved from Cambridge to McGill University in Montreal, Canada, he estimated the amount of energy being released from the uranium ore (based on its alpha-ray emissions alone), and concluded – in spite of under-estimating the figure by a considerable margin – that 'it is difficult to suppose that such a quantity of energy can be derived from regrouping of the atoms or molecular recombinations on the ordinary chemical theory'. This was a soberly phrased way of saying that something utterly fundamental was missing from the current understanding of the nature of matter.

At McGill, Rutherford began his immensely productive but difficult collaboration with Soddy. Radioactivity was at that time still very much in the province of chemists, and Rutherford did not know much about chemistry. In fact, the phenomenon of radioactivity seemed to insist on an entirely new form of chemistry, because researchers started to find new substances that appeared almost by magic in radioactive materials. As we saw in the previous chapter, Rutherford, in collaboration with R. B. Owens, discovered a kind of gas or 'emanation' coming from thorium compounds, which he and Soddy later christened thoron. They also found what seemed to be a new element in these materials, which they called thorium-X, by analogy with the so-called uranium-X that the British chemist William Crookes claimed to have discovered as an impurity in uranium salts. It looked as though thorium-X was produced from thorium itself, and that thorium-X, rather than thorium, was the true source of thoron. In other words, thorium was apparently *turning into* thorium-X, through what the two researchers called, with coyly worded heresy (and without really knowing what it meant), 'sub-atomic chemical change'.

It was hard to avoid the conclusion that thorium, and uranium and radium too, were indeed disintegrating and becoming transformed into new elements. Rutherford realised that this disappearance of 'parent elements' and the accumulation of 'daughter elements' could be used as a kind of clock to estimate the ages of minerals. He and Soddy showed that radioactive decay obeyed a precise mathematical law: the amount of radiation coming from a sample declined exponentially over time. This means that it falls to half its initial level in a certain time period, and then takes the same time to fall from a half to a quarter, and from a quarter to an eighth, and so on. This characteristic time period is called the half-life. If the half-life of a radioactive element is known,* the amounts of it and its daughter elements in a rock can be used to estimate the time when this radioactive 'clock' started ticking: the age of the rock, in other words. Rutherford's colleague, the Yale chemist Bertram Boltwood, used this principle in 1907 to deduce that the earth is at least two billion years old: about 20 times greater than the previous estimate made by Lord Kelvin in the 1860s.

How was it was possible for one element to be transformed into another? This was clearly related in some way to the nature of the 'rays' that were emitted when the transformation took place; but the significance of that relationship was by no means obvious. At first, a common view was that alpha-rays were secondary phenomena produced when beta-rays (electrons) released during radioactive decay struck the parent material, just as the X-rays seen by Roentgen were emitted from the glass of a cathode-ray tube when it was hit by the stream of electrons. But Rutherford and Soddy found that when they separated their thorium-X from thorium it carried away with it the beta radiation: the parent thorium emitted only alpha rays. Thus the decay of thorium to thorium-X appeared to be fundamentally connected to the production of alpha rays.

Thomson had deduced that cathode rays were negatively charged particles by looking at how they were deflected in electric and magnetic fields. By measuring this deflection, he was able to work out the ratio of the electrical charge on each particle to its mass: the so-called *e/m* ratio. Researchers had naturally looked to see whether alpha rays could be deflected in this way too, but they had failed to observe it. Rutherford approached the question anew with his characteristic experimental acumen and dexterity. He made a small metal box open at each end, like the outer shell of a matchbox, and then divided it up into a series of narrow channels. He blocked part of the end of each channel with a metal strip. Then he charged up the channels with an electric field and fired alpha rays down them. If the rays were deviated from their course, some of them would fall below the metal strip at the end and fail to exit, resulting in a decrease in the intensity of radiation leaving the box. To investigate the effect of a magnetic field, he placed his little device between the poles of a powerful electromagnet.

---

* Strictly speaking, we cannot talk about the half-life of a particular element, but only of a specific 'atomic form' or *isotope* of that element. Different isotopes have slightly different nuclear masses – we will see why in the next chapter – even though all isotopes of an element have the same number of protons in their atoms, and essentially the same chemical behaviour.

This revealed to Rutherford that alpha rays were indeed electrically charged particles – and remarkably hefty ones, in comparison with the nimble electron. They had a positive charge, and if one assumed that this was the same order of magnitude as the charge on an electron, the particles had to be about as massive as an entire hydrogen atom. But that made sense. For if radioactive atoms spit out atom-sized particles, then it was more understandable that they could become converted into entirely new types of atom. Radioactivity, in other words, was not the *result* of transmutation, but the *cause* of it. 'This result', said Rutherford to J. J. Thomson, 'is of great importance in its application to radioactive bodies as it gives a mental picture of what is taking place in the succession of changes that is going on.'

The *e/m* ratio of the alpha particles seemed to point to just two possibilities. If they carried an equal but opposite charge to that of the electron, then they had half the mass of a helium atom. But the same *e/m* ratio would result if they had the mass of a whole helium atom, but *twice* the charge of an electron. There was no way to distinguish between these possibilities. 'There is very little actual evidence at the present time', Soddy wrote in 1906, 'on which to base a conclusion.'

The gossamer tube of Mr Baumbach

According to the Hungarian chemist Georg von Hevesy, there were only two men Rutherford did not like: Frederick Soddy and William Ramsay. Before they started working together at McGill, Rutherford and Soddy argued heatedly, and publicly, over whether atoms were divisible (to a chemist like Soddy, that was a highly unattractive notion). Rutherford's clashes with Ramsay began when Soddy left Canada to work with the discoverer of the noble gases in London, and it must have seemed to Rutherford in his darker moments that the two men were plotting against him.

Rutherford harboured the strong suspicion that alpha particles were helium atoms, but he could not prove it. In 1903, Soddy and Ramsay conducted an experiment that Rutherford had initially proposed to his former colleague, in which a radium salt was placed in a glass vacuum tube and any gas produced by the radioactive decay was collected and analysed. Ramsay and Soddy collected a tiny bubble, which spectral analysis (see page 44) identified as helium. This, however, did not prove that alpha particles were themselves helium, and the two chemists did not make that leap of reasoning – maybe the helium was produced by the action of radiation on the salt. Ramsay began speaking about his results without any acknowledgement of Rutherford's part in conceiving the experiment, which led J. J. Thomson to write indignantly from England to his former student in Canada:

> I have been utterly disgusted with Ramsay's behaviour and have said so on many occasions. I always, when speaking on the subject, point out how you had foreseen and indeed planned the experiments, and that if it had not been for you they would never have been made. It is a great pity his morals are so inferior to his manipulation.

Rutherford himself seems to have shrugged off such disputes, allowing himself only the occasional jocular reference to the 'Ramsamania' of the English duo. Yet Ramsay's work must only have hardened Rutherford's resolve to get at the identity of 'his' alpha particle. He was still struggling with the problem when he left McGill in 1907 and returned to England to work at Manchester University. 'The determination of the true character of the alpha-particle is one of the most pressing unsolved problems in radioactivity, for a number of important consequences follow from its solution', he said in January 1908.

At Manchester Rutherford acquired one of his most able students, Hans Geiger, and together they worked at developing an instrument that could detect alpha particles one by one. This involved firing the particles into a gas at low pressure between electrodes charged with a strong electric field. Each particle could produce around 80,000 ions in the gas. That created an electrical discharge across the electrodes, which signalled the passage of the alpha particle. This instrument was the forerunner of the Geiger counter, the standard device for measuring radioactivity.

Knowing the alpha particle's ratio of charge to mass (e/m), one could calculate the mass (and thus the chemical identity of these atom-sized particles) if only one could measure how much charge it carried. This is what Rutherford and Geiger did in 1908. Rutherford had already estimated the total charge emitted by a quantity of radium while he was in Canada, and by now counting how many alpha particles this corresponded to, he and Geiger were able to conclude that indeed the particles have the same atomic mass as helium. 'An alpha-particle', they wrote, 'is a helium atom, or to be more precise, the alpha-particle, after it has lost its positive charge, is a helium atom.'

But just as Marie Curie knew that her claim that radium was a new element would not be accepted by chemists until she had isolated it in pure form, Rutherford recognized that arguments based on numbers would not be as compelling as a demonstration that alpha particles and helium were chemically the same thing. Working with his student T. Royds, he began planning his masterly experiment in the autumn, around the beginning of the 1908–9 academic year.

It required an apparatus as delicate and remarkable as it was simple. Although Rutherford enjoyed devising and making his own instruments, the task he now faced was beyond his practical skills. He sought the help of a glassblower named Otto Baumbach, who ran his business from premises near the university. Baumbach made for Rutherford a glass capillary tube of incredible delicacy: the walls were made of a glass membrane just one hundredth of a millimetre thick. Even Baumbach couldn't achieve such precision easily. 'After some trials', Rutherford wrote, 'Mr Baumbach succeeded in blowing such fine tubes very uniform in thickness.'

He then enclosed this tube in a larger, stouter tube fitted with taps to allow it to be evacuated or filled with other gases. The central capillary was filled with radium, and the outer tube was pumped free of gases. Rutherford figured that, if the wall of the inner tube was thin enough, alpha particles ejected from the radioactive source would be able to pass straight through it, so that they – and they alone – would be captured in the outer tube (Figure 5). If Rutherford was right, then helium gas would gradually accumulate in the outer tube – as a measurement of the spectrum of the gas would verify.

But, one might object, is this so different from the experiment of Ramsay and Soddy, who had already shown that helium was emitted from radium? The difference is that if helium is just a *by-product* of radioactive decay it is released as an ordinary gas, the atoms of which are not fired like bullets but merely diffuse like drifting motes of dust. Rutherford and Royds showed that in that case helium atoms were not able to penetrate the thin glass wall: if they filled up the inner tube with helium itself, none escaped through the glass. The only way the helium could get through was in the form of energetic alpha particles.

The experiment is so simple in concept that it is easy to overlook what a profound issue it is probing. The skills of a Manchester glassblower were permitting Rutherford to look *inside the atom*, to catch a glimpse of what happens when one chemical element is transmuted spontaneously into another. To effect such a transmutation, the

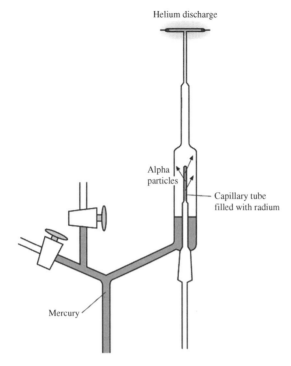

Figure 5 *The apparatus that enabled Rutherford to demonstrate that alpha particles are helium ions. The narrow capillary has walls so thin that energetic alpha particles, emitted from a radium compound, can pass straight through, becoming trapped in the outer tube. Here they are studied spectroscopically to reveal their chemical identity: an electrical discharge between two electrodes in the topmost part of the tube causes the helium gas formed from the alpha particles to glow with a characteristic spectral signature*

atom itself must fall apart. The identity of the alpha particle is the crucial clue to the nature of that disintegration. If it is indeed a helium atom, then it would appear that big atoms are made up from smaller atoms: that there is some genetic relationship between the elements. That had long been suspected – the British chemist William Prout proposed in 1815 that atoms of all the elements are conglomerates of hydrogen atoms, since hydrogen is the lightest of all elements. Prout's hypothesis had some respectable supporters, but it was opposed by authoritative heavyweights such as Jöns Jacob Berzelius and Dmitri Mendeleyev. The astronomer Norman Lockyer, who discovered helium in 1868 by identifying a new spectral signature in sunlight (the name comes from the Greek *helios*, Sun), elaborated on Prout's idea by suggesting in the 1870s that the elements *evolve* from one to another within the furnaces at the hearts of stars. William Crookes – who, you will recall, claimed that uranium decomposes via radioactive decay into a new substance he called uranium-X – concocted a similar theory of stellar transmutation. Radioactivity had, since the work of the Curies, seemed to support this kind of transformation of elements; but an identification of the alpha particle with helium would provide concrete evidence that atoms are composite particles, fragments of which have their own distinctive chemical nature.

Rutherford announced the results of his experiment on 3 November 1908 at an address to the Manchester Literary and Philosophical Society, entitled simply 'The nature of the alpha-particle.' The experiment worked just as planned, he told the assembly, and the spectral measurements verified the presence of helium in the outer tube. If this seems like a somewhat parochial forum in which to disclose a major scientific discovery, it is not: the Manchester society had a distinguished pedigree. It was here in 1803 that the Society's secretary John Dalton publicly unveiled his atomic theory – a fact surely not lost on Rutherford, who chose the same platform to announce his later epoch-making discovery of the atom's internal structure.

Yes indeed – there were still greater things to come from Rutherford's days at Manchester. But by 1908 he was already internationally renowned for his work on the nature of radioactive decay, and that was recognized at the end of the year when the message came from Sweden that Rutherford had been awarded the Nobel prize 'for his investigations into the disintegration of the elements, and the chemistry of radioactive substances'. To his immense amusement, the prize was awarded in chemistry, not physics: the Nobel committee in chemistry, concerned that the Curies and Becquerel had been awarded the 1903 prize in physics, were keen to reclaim this exciting area of science for themselves. This disagreement over whether radioactivity and atomic science belonged to physics or chemistry reflects the fact that these topics do indeed define the meeting point of the two disciplines. For Rutherford, however, who felt that he spent his life battling against the reluctance of chemists to see their precious atoms dissected, the 1908 prize was a grand irony.

His address to the Stockholm grandees in December of that year left no doubt where his priorities lay. The award was not, of course, for his latest work on the alpha particle, which no one knew about when the decision was made – but that was nonetheless the subject of Rutherford's Nobel lecture, entitled 'The chemical nature of the alpha particles from radioactive substances.' Here he described the experiment that he and Royds had conducted just weeks earlier. 'It is very remarkable', he concluded, 'that a chemically inert element like helium should play such a prominent part in the constitution of the atomic systems of uranium and thorium and radium . . . It is consequently not unreasonable to suppose that other elements may be built up in part from helium.'

Inside the atom

His obsession sated, Rutherford went on to use the alpha particle as his principal tool for probing the atom. He fired them like bullets at thin metal films, and found that most of the particles passed straight through, in line with the findings of Philipp Lenard in 1903 that cathode rays (electrons) can pass unhindered through matter. Rutherford's alpha-particle bombardments at Manchester persuaded him that atoms were not like Dalton's hard, massy spheres, but were tenuous bodies that were mostly empty space. Yet in his most famous experiment, conducted in 1909 with Geiger and fellow student Ernest Marsden, he shot a stream of alpha particles at a thin foil of gold leaf and found to his astonishment that some of them bounced right back – 'as if you had fired a 15-inch shell at a piece of tissue paper and it came back and hit you'. The atom, he decided, was not the 'plum pudding' suggested by J. J. Thomson, a blob of matter with a positive electrical charge in which negative electrons were embedded. Instead, it contained a tiny nucleus of positive* charge, where just about all of the mass resided, surrounded by a thin cloud of electrons. In a letter to the Japanese physicist Hantaro Nagaoka in 1911, Rutherford explained that

> I have devised an atom which consists of a central charge 'ne' [that is, some integer n times the charge e on an electron] surrounded by a uniform spherical distribution of opposite electricity, which may be supposed, if necessary, to extend over a region comparable with the radius of the atom as ordinarily understood.

---

* It was not immediately obvious that the nuclear charge had to be positive; indeed, at first Rutherford thought it was negative, and that alpha particles executed comet-like arcs around this nucleus.

This letter was by way of acknowledgement that, as Rutherford discovered only that year, Nagaoka had proposed something similar in 1903: a 'Saturnian' atom in which electrons orbited a positive core like the rings around Saturn. But Nagaoka's nucleus was by no means tiny, whereas Rutherford's experiments on the scattering of alpha particles convinced him that the nucleus was a mere dot in the centre of an atomic void.

That, of course, was just the beginning of a revolution – one that led to Rutherford's artificially induced splitting of the atom in 1919, to the realization that this 'nuclear fission' could be an awesome source of energy, to the bright promise of nuclear power and the grim reality of nuclear warfare. It led also to a new kind of physics, thanks largely to a Danish scientist who, Rutherford wrote from Manchester in 1912, 'has pulled out of Cambridge and turned up here to get some experience in radioactive work'. His name was Niels Bohr. Bohr's quantum atom, in which the circuits of the electrons around the nucleus were confined to specific, quantized orbits, led to an understanding of how the disposition of electrons in atoms gives rise to the unique chemical properties that distinguish one element from another, rationalizing the familiar twin-towered arrangement of the periodic table.

CHAPTER 5

The Elements Came in One by One

Seaborgium's Chemistry: Small is Beautiful

Darmstadt, 1995–7—An international team of scientists use a particle accelerator to fuse two atoms together, creating an artificial, 'superheavy' element named seaborgium. This is not the first time the element has been made, but the researchers are now equipped to examine its chemical properties. There are just two problems: they can make only about one atom an hour, and the atoms last for just a few seconds before radioactively decaying. The feat of conducting chemistry one atom at a time with ephemeral seaborgium stretches the techniques of chemical analysis about as far as they will go.

There are machines for making new elements. Only a handful of them exist, but they are gradually extending the periodic table along its bottom margin, where the elements are made from relatively immense, ponderous atoms, much heavier than lead. These substances cannot be found in nature; they are the products of nuclear reactions in which atoms are brought crashing into one another with great energy. Occasionally these atoms stick together: there is nuclear fusion, an artificial and in some sense exaggerated version of the process that happens inside stars to generate the raw stuff of the universe.

Extremely heavy atoms tend to fall apart very easily. They are liable to undergo radioactive decay, spitting out Rutherford's alpha particles. Alternatively they may rupture into two roughly equal fragments: this is nuclear fission, and the energy it releases is harnessed in the nuclear reactors used today for power generation.

Whereas the elements throughout most of the periodic table are composed mainly of atoms that remain stable indefinitely, the heavy elements at the bottom of the table tend to consist of unstable, potentially fissile atoms. Some, such as the isotopes (see below) of uranium that occur in pitchblende, decay only slowly, over millions or even billions of years. Others decay faster. The most durable atoms of einsteinium, element number 99 in the sequential list that begins with hydrogen, typically decay within a year and a half.* Einsteinium was not known until it was found in the fallout from early hydrogen-bomb testing in 1952.

But the elements made in the current generation of atom factories are far more ephemeral than einsteinium. They survive for perhaps a few seconds; some have half-lives measured in milliseconds. And there's the rub. To chemists, making new elements is only the start of the matter. For the story of chemistry is all about how elements behave – how they react with one another to form combinations of elements, how they may be shuffled and grouped into symphonies of atoms. But how do you investigate the chemical behaviour of atoms that exist for only a few seconds?

The answer is that you devise an experiment that, in its technical aspects at least, is quite beautiful.

Making elements

The first new human-made element was not particularly heavy. Called technetium, it is number 43 in the sequence, lodged between molybdenum and ruthenium in the periodic table, and is something of an anomaly among these moderately heavy elements in that it has no long-lived atomic forms.

What do I mean by 'atomic forms'? In the previous chapter I introduced these as *isotopes*, but without quite saying what isotopes are. Atoms are composed of a dense nucleus made up of two kinds of subatomic particle – protons and neutrons – surrounded by a cloud of electrons. The number of electrons determines the chemical

---

* As we saw on page 58, the decay time of a radioactive substance is measured in terms of its *half-life*, which is the time taken for half of the atoms in any given sample to decay. The half-life of the longest-lived isotope of einsteinium, denoted einsteinium-252, is 470 days. So if you were to be given a gram of einsteinium-252 (and in fact it can't even be made in microgram quantities yet), then in 470 days time you'd have only half a gram left. The rest will have decayed to other heavy elements, which would in turn decay at a rate determined by their own half-lives.

behaviour of the element,* and this in turn depends on the number of protons: electrons and protons have equal but opposite electrical charge, and in a neutral atom there is an equal number of each. The number of protons in an element is called its *atomic number*, and it is this number that defines the ordering of elements, reading from left to right and down successive rows in the periodic table.

Now, I mentioned earlier that different isotopes of an element have the same number of protons in their nuclei – and thus the same chemical properties – but different atomic *masses*. The reason for this is that they differ in the number of neutrons (which are electrically neutral) in their nuclei. For lighter elements, the nuclei have roughly the same number of neutrons and protons – but the numbers do not have to be *exactly* the same. For example, while the most common isotope of carbon, called carbon-12 and denoted ^{12}C, has six protons and six neutrons, radiocarbon dating relies on the isotope carbon-14 (^{14}C), which has eight neutrons per atom. In other words, its *atomic mass* – the total number of protons and neutrons – is 14. The nucleus of carbon-14 is unstable: it undergoes radioactive decay, with a half-life of about 5730 years, by emitting a beta particle. That's why it is sometimes called radiocarbon. Radiocarbon dating is a way of estimating the age of an object made from organic material (that is, carbon-based material derived from a living organism) by measuring how much of its radiocarbon has decayed. Carbon-14 is constantly being formed in the atmosphere by the collision of cosmic rays with molecules in the air, and while an organism is alive it is constantly taking in 'fresh' radiocarbon. When it dies, this renewal stops and the radiocarbon clock starts ticking.

So it's not just heavy atoms that are susceptible to radioactive decay – some isotopes of light atoms decay too, generally because they have 'too many' or 'too few' neutrons. There is at least one other radio-active isotope of carbon, ^{11}C, as well as a second stable isotope, ^{13}C, which constitutes about 1 percent of all the carbon in the world.

The curious thing about technetium (chemical symbol Tc) is that it has no isotopes that are not radioactive, nor even any that have very long half-lives. The longest-lived isotope of Tc has a half-life of about four million years – which may sound like a long time, but is not enough to preserve any significant amounts of technetium that might have been incorporated into the Earth when it formed over 4.5 billion

* More precisely, this behaviour depends on how many electrons the atom has *and* how they are arranged in shells and subshells – see pages 141–2.

years ago. Thus, there is almost no technetium in our natural environment.* Yet scientists knew such an element must exist, because it left a hole in the periodic table: there were elements up to number 42 (molybdenum) and from number 44 (ruthenium), so there must surely be an element 43. Chemists searched for it for decades, and there were several false claims (accompanied by fanciful names) of identifications of element 43 in minerals. It was not until 1937 that the first convincing evidence for element 43 was reported.

By firing alpha particles at atoms, Ernest Rutherford had found that he could induce radioactive decay of formerly stable atoms, transmuting one element to another. In 1919 he showed that nitrogen could be transmuted this way – his celebrated 'splitting of the atom'. This kind of modern alchemy was invigorated by the invention of the particle accelerator. In 1929 the American physicist Ernest Lawrence at the University of California at Berkeley came across an article by a Norwegian engineer named Rolf Wideröe that explained how electrically charged plates could be used to accelerate positively charged particles (ions such as alpha particles) to high energies. Lawrence had been searching for a way to do this, as it would make such particles better able to penetrate into atomic nuclei and cause transmutation. At the Cavendish laboratory in Cambridge, John Cockroft and Ernest Walton had the same idea. Cockcroft and Walton made a linear accelerator, in which particles were boosted along a straight-line trajectory; but Lawrence saw that longer paths, allowing for greater acceleration, could be accommodated by devising plates that sent the particles into a spiral trajectory. He called his device a cyclotron.

Researchers at Berkeley used Lawrence's cyclotron in 1937 to bombard a thin foil of molybdenum metal with small particles: not alpha particles but the nuclei of 'heavy hydrogen', the isotope hydrogen-2, which is also called deuterium. These deuterium nuclei, or deuterons, contain one proton and one neutron. The irradiated foil was shipped across the Atlantic to the University of Palermo in Sicily, where chemists Carlo Perrier and Emilio Segrè analysed it and discovered two previously unknown radioactive substances, with half-lives of 62 and 90 days. These turned out to be two isotopes of the missing element 43, which ten years later the Italian researchers named technetium after the Greek word *technetos*, meaning 'artificial'– or as Francis Bacon might have said, the product of 'art'.

---

* Tiny amounts of this element are constantly produced by radioactive decay of other elements, and by the bombardment of other elements by cosmic rays.

Segrè, whose skills at chemical analysis made him one of the great pioneers of radiochemistry, was involved in the discovery of a second 'missing element', number 85 in the periodic table. Named astatine after the Greek word for 'unstable', it was made in 1940 by Segrè working with the Americans Dale Corson and Kenneth MacKenzie at the Berkeley cyclotron. (Segrè was by that time a fugitive from Mussolini's Italy, and Berkeley was the obvious refuge for him.) The researchers used the accelerator to fire alpha particles at bismuth (element 83). An alpha particle that fuses with a bismuth nucleus adds two more protons to produce element 85.

The Berkeley team made the isotope astatine-211, which has a half-life of just over seven hours. The time window for exploring the chemistry of these human-made elements was apparently getting smaller.

But Segrè wasn't only interested in filling in the gaps of the periodic table. He wondered whether the table itself could be extended. As things stood in the 1930s, the table ended at element 92, which was uranium. Could atoms heavier than those of uranium be made by bombarding this ponderous metal with subatomic particles in the hope that some might stick? Segrè began to investigate that question in 1934, working in Rome in collaboration with one of the genuine giants of nuclear physics, Enrico Fermi. The neutron had itself only been discovered two years previously, by James Chadwick in England, and the Italian researchers began looking for new elements – which they hoped to identify by their distinctive radioactive decay signatures – by firing neutrons at uranium. In 1934 Fermi and his colleague Oscar D'Agostino tentatively reported the existence of two new elements heavier than uranium ('transuranic' elements), with atomic numbers 93 and 94. But this claim was later shown to be false.

Five years later Segrè was at Berkeley and working with physicist Edwin McMillan, still shooting neutrons at uranium. They found what they initially thought was a new element, temporarily dubbed eka-rhenium. The Sanskrit prefix 'eka' was used by Dmitri Mendeleyev to refer to 'missing' elements in his periodic table by reference to their presumed chemical similarities with other known elements: eka-rhenium was expected to resemble rhenium. But at length Segrè and McMillan concluded that eka-rhenium was just one of the elements known already: one of the series of exotic metals called the lanthanides.

They were wrong, however, as McMillan discovered when he was joined in 1940 by chemist Philip Abelson. With Abelson's expertise, eka-rhenium did indeed prove to be a new element, number 93, which, by analogy with the naming of uranium after the planet Uranus, they decided to call neptunium (Np).

Official secrets

The war changed everything. It was clear to nuclear physicists and chemists that nuclear fission of a heavy element like uranium could provide an enormous, almost inexhaustible source of energy. But how to release it? As Rutherford had shown, radioactive decay could be *induced*. In 1934 the Hungarian physicist Leo Szilard realised that if, on fission, an atom emitted particles (such as neutrons) that induced the fissile decay of other atoms, this could lead to a chain reaction in which the release of nuclear energy happened in all the atoms almost simultaneously. All you needed was enough of the material to get the chain reaction going – a critical mass.

The idea became concrete in 1940, when Soviet physicists Konstantin Petrzhak and George Flerov discovered that the atoms of one isotope of uranium, ^{235}U, undergo spontaneous fission. The process also releases neutrons that can induce fission of other uranium atoms, raising the possibility of a chain reaction. But the catch was that ^{235}U is a relatively rare isotope, constituting just one percent of natural uranium. To create a fission bomb, this isotope would have to be separated from the more abundant ^{238}U, with which it was chemically all but identical. That became one of the major challenges for the Manhattan Project, convened in 1942 to deliver America the atom bomb.

The military potential of nuclear research was already apparent by 1941, and that was why the discovery in that year of the next artificial element after neptunium wasn't reported until 1946. It was made at Berkeley by a team led by the young chemist Glenn Seaborg (Figure 6), by bombarding uranium with deuterons. This process first produced a hitherto unknown isotope of neptunium, ^{238}Np, which decayed by beta emission to form a new element, number 94. After Uranus and Neptune lies Pluto, the last planet in the solar system; and so element 94 became plutonium, whose chemical symbol Pu contained an element of undergraduate humour. The Berkeley team, joined by Segrè, subsequently made another isotope of plutonium, ^{239}Pu, which was fissile in the same way as ^{235}U. This was the stuff that exploded over Nagasaki four years later.

Figure 6 *Glenn Seaborg, a pioneer of element-making and the only person to have had an element named after them while they were still living. Here he points to seaborgium in the periodic table*
(© CORBIS)

Plutonium was the key to the next two artificial elements, which Seaborg and his Berkeley colleagues created in 1944. That summer they irradiated plutonium-239 with alpha particles to make element 96 (called curium, honouring Marie and Pierre Curie), and later in the year they made element 95 (americium) by bombarding plutonium with neutrons in a nuclear reactor. Both of these human-made elements have practical applications. Americium-241 is used in smoke detectors, where the alpha particles emitted as the element decays ionize the air and allow a current to flow between two electrodes – just as in a Geiger counter. Smoke particles soak up the ionized air molecules and cause the current to drop. Curium is used as a power source for very-low-power electrical devices that aren't easily accessible, such as pacemakers and navigation buoys.

What were the chemical properties of these new elements? The central feature of the periodic table is the existence of groups of elements that show comparable properties. It is called the *periodic* table because certain chemical properties recur periodically as one progresses through the list of increasing atomic number from 1 (hydrogen) to 92 (uranium). Thus, lithium behaves similarly to sodium and potassium; fluorine forms compounds like those of chlorine and bromine. These 'chemical families' were noticed in the early part of the nineteenth century, but it wasn't until 1869 that Mendeleyev figured out how to arrange the elements in a table that captured all the patterns of behaviour (other chemists had already drawn up rather similar tables a few years previously).

While the atomic numbers of the elements increase steadily across each row of the periodic table, the vertical columns correspond to the chemical families. The column on the far left, for example (group 1), contains the alkali metals, including lithium, sodium and potassium. One can predict some of the chemical properties of an element, such as how many chemical bonds it tends to form (its valency; see page 141), from those of the other elements in the group. By this token, the elements that, from 1940, began to extend in a new row beyond thorium and uranium were expected to behave like those that apparently sat above them: plutonium like osmium, americium like iridium, curium like platinum and so on (Figure 7a).

Now, it turns out that plutonium is not unlike osmium, just as uranium forms some compounds that are similar to those of tungsten. But these correspondences seemed to break down for americium and curium: for example, the formula of platinum oxide is PtO_2, whereas curium forms an oxide with formula Cm_2O_3. Elements 95 and 96 seemed more closely to resemble the metals in the so-called lanthanide series: a group of 14 elements, from 58 (cerium) to 71 (lutetium), that is squeezed into the periodic table between lanthanum and hafnium.

Indeed, this unexpected chemical behaviour foiled early attempts to identify the two new elements, because it misled researchers trying to separate them from other elements. They weren't isolated until Glenn Seaborg had the crucial insight that perhaps they were part of a series analogous to the lanthanides, commencing with thorium (which comes after actinium). In other words, this series of ultraheavy elements comes 'unstuck' from the row above it, looping out like the lanthanides into a distinct family of its own. Once americium and curium were found to confirm Seaborg's hunch, he proposed calling this group of elements the *actinides* (Figure 7b).

Seaborg's intuition was borne out as the Berkeley team continued to extend the table, using each new element as the raw material

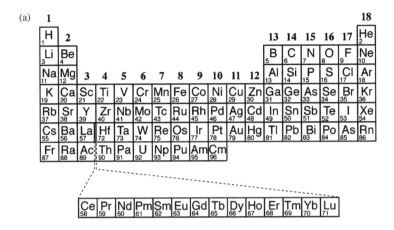

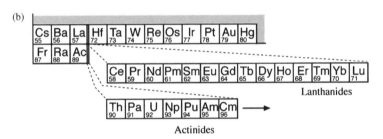

Figure 7 *In the periodic table as it seemed to stand after the discovery of elements 95 and 96 (americium and curium), these two elements appear to sit in the same columns as iridium and platinum respectively (a). But squeezed in the row above is the series known as the lanthanides; and Glenn Seaborg guessed that the elements after actinium constitute an analogous series, called the actinides (b). Here the atomic numbers of the elements are indicated in the lower left of each box. The groups of elements in each column are conventionally numbered as shown*

for another further along the row. In 1949 they made element 97 (berkelium, what else?) by blasting americium-241 with alpha particles; and the same process applied to curium-242 in 1950 gave them element 98 (californium, more or less exhausting the geographic options). The five-year hiatus in the manufacture of new elements since 1944 was the consequence of having to patiently create and purify enough of the preceding elements to provide a target – curium-242 has a half-life of just 162 days.

Berkelium and californium are even more fleeting: the isotopes first prepared by the Berkeley team had half-lives of four and a half hours and 44 minutes, respectively. But as I intimated earlier, the subsequent element (number 99, einsteinium), as well as element 100 (fermium), required no such painstaking preparation. They both turned up in the debris collected on a Pacific atoll next to Eniwetok in the Marshall Islands, where a prototype bomb dubbed 'Mike' unleashed the power of the sun by triggering the nuclear fusion of hydrogen-2 and hydrogen-3 (deuterium and tritium). Just as the fission of heavy atoms releases nuclear energy, so does the *fusion* of light atoms: this is the process that powers stars, and it results in the production of successively heavier atoms from hydrogen and helium. But to create the conditions needed to ignite fusion in Mike's hydrogen core, a fission bomb made from uranium acted as the fuse. In the process, some of this uranium was transformed into transuranic elements, including einsteinium and fermium.

Scientists at Berkeley subsequently found that they could make einsteinium in a nuclear reactor by irradiating plutonium with neutrons. This allowed them in 1955 to collect enough of the artificial element – about a billion atoms – to use it as a target for alpha particles, with the aim of producing element 101 (later called mendelevium). The researchers estimated that from a billion einsteinium atoms they could expect to produce precisely one atom of element 101.

How on earth do you spot one atom among a billion others? In principle, chemical techniques for separating these metallic elements were well established. The most important is called ion-exchange chromatography, which involves passing a solution containing ions of the various elements through a column packed with resin-coated beads. To separate metals, the resin is typically a polymer containing negatively charged chemical groups that bind to the positively charged metal ions. The different types of ion stick to the resin with different degrees of 'stickiness', depending for example on differences in their size or their electrical charge. They can be removed again by 'eluting' or washing the beads with another liquid that redissolves the ions. The least-tightly bound are removed first; and so each element is flushed out of the column in a separate pulse (Figure 8a). Ion-exchange chromatography is used, for example, to separate the various members of the lanthanide family of metals, which are chemically very similar (Figure 8b).

All very well; but can you see a single atom in the eluting fluid? Well, you can if it is radioactive with a short half-life, because radiation detectors can pick up the particle that is emitted when the atom decays: sufficiently sensitive detectors can see every single one of these particles. Using this method, one night in February of 1955 the Berkeley team

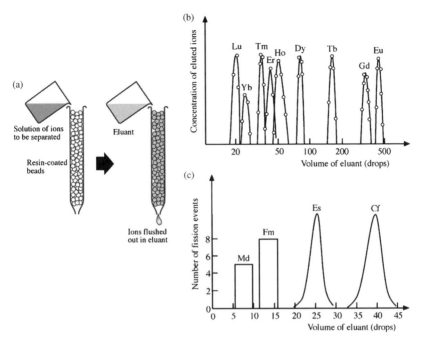

Figure 8 *In ion-exchange chromatography, chemically similar elements are released from a column packed with resin beads at different times in the 'elution' procedure (a). This is how the lanthanide elements may be separated: each element comes out of the column in a more or less separate pulse (b). The technique allowed just a few atoms of element 101 (mendelevium, Md) to be separated from other actinides (c)*

analysed the material produced by alpha irradiation of their billion or so atoms of einsteinium, and detected the radioactive-decay signatures of precisely 17 atoms in the elution liquid from the ion-exchange column, which they could ascribe to spontaneous nuclear fission of the new element 101 (Figure 8c). The isotope produced this way has a half-life of just over an hour. This, say Seaborg and his colleague Walter Loveland of Oregon State University, 'was the first case in which a new element was produced and identified one atom at a time.'

Putting on more weight

Now you might begin to see what the element-hunters are up against. Although there was no simple relationship between the atomic mass of these new artificial elements and their half-life, roughly speaking it is fair to say that the heavier they got, the more quickly they decayed. On top of that, they became increasingly hard to make. Four transuranic elements were made between 1941 and 1944; finding the next five took 11 years. Element 102 was made in 1958 (it had already acquired the name nobelium after an earlier false claim) and 103 (lawrencium) in 1961. But already there were disputes. With the Cold War nuclear programs in full swing, the discovery of new elements became a matter of nationalistic pride – a symbolic demonstration of technological superiority – and the Berkeley group found themselves in competition with a Soviet team at the Joint Institute for Nuclear Research in Dubna, who claimed to have made element 102 in 1956. In 1964 they said they had detected an isotope of element 104 with a half-life of 0.3 seconds. The Berkeley group denied that any such isotope existed when, in 1969, they produced their own version of element 104 and named it after the man who split the atom: rutherfordium (Rf).

The same fate befell element 105: claimed by the Dubna team in 1967 (who called it, rather clumsily, nielsbohrium) but disputed by the Berkeley team in 1970 (they wanted to call their version hahnium, after the German scientist Otto Hahn who, with Fritz Strassman and Lise Meitner, discovered nuclear fission). These squabbles led to much argument and confusion about what the new elements should be called, and it wasn't until 1997 that the International Union of Pure and Applied Chemistry (IUPAC) – the official authority on chemical nomenclature – ruled that 104 was indeed to be known as rutherfordium, while the precedence of the Dubna team for element 105 was acknowledged by the name dubnium (Db).

IUPAC was also called upon to adjudicate over rival claims to the discovery of elements 106 and 107. The Dubna group, led by Yuri Oganessian, asserted that they had made the first of these in 1974 and the second in 1976. The first claim was disputed by American researchers at Berkeley (then headed by Albert Ghiorso, who had worked with Seaborg since the curium synthesis in 1944) and the Lawrence Livermore National Laboratory. The Russian team's claim to element 107, meanwhile, was questioned by a new contestant in the race for new elements, a German team at the Laboratory for Heavy Ion Research (Gesellschaft für Schwerionenforschung; GSI) in Darmstadt, who had obtained more convincing results in 1981.

Whereas previously new elements had been created by bombarding heavy atoms with small, light ones (such as deuterons or alpha particles, which are of course helium nuclei), the GSI team perfected a different approach that was to put them at the head of the race throughout the 1980s and 1990s. They brought about the fusion of two medium-sized nuclei by firing ions of metals such as zinc, nickel or chromium at targets made of relatively dense metals like lead and bismuth. The method was developed in the 1970s by the Dubna team, who used it to make fermium and rutherfordium and in their claimed syntheses of elements 106 and 107. But in the hands of the GSI team, the technique generated all the elements from 107 to 112, the last of these in 1996. The isotope of 112 with a mass of 277, made at GSI by bombarding a lead target with zinc-70 ions, has a half-life of about a quarter of a millisecond.

In 1993, IUPAC ruled that the American claim to element 106 was more convincing than that of the Dubna group, and this left the American team at liberty to choose a name. They decided to honour Glenn Seaborg, then the elderly statesman of nuclear chemistry, by calling 106 seaborgium (Sg).

But IUPAC did not like that at all, for Seaborg was still alive (he died in 1999). No element had previously been named after a living person. But when IUPAC rejected the proposed name, the American Chemical Society rebelled, declaring in 1995 that all of its journals would henceforth refer to element 106 as seaborgium; and IUPAC was forced to relent. As for element 107, its discovery was awarded jointly to the GSI and Dubna teams, who agreed to name it nielsbohrium, which was shortened to bohrium (Bh).

Elements 108 and 109 are now called hassium (after Hesse, the German state in which GSI is situated) and meitnerium, after Lise Meitner. Happily, the furious competition in element-making has now given way to international collaboration, so that elements 114 and 116, made in 1999 and 2000 respectively, were the result of a joint effort between the Dubna and Lawrence Livermore groups. The same collaboration reported the synthesis of element 115, and its decay into element 113, in early 2004, and later that year a Japanese team led by Kosuke Morita reported the synthesis of element 113 at the Accelerator Research Facility of the Japanese research organization RIKEN. The quest continues, with element 118 the next target in line.*

Nuclear scientists are particularly excited by the notion that there may be isotopes of these 'superheavy' elements, with atomic number around 114, that are particularly stable, so that unlike the other superheavies they do not decay virtually as soon as they are formed.

* The synthesis of element 118 was claimed in 1999 by the Berkeley group, but was subsequently retracted. The Dubna team claimed to have spotted two nuclear-decay events of the sort expected for element 118 in 2002, but this result has not been confirmed by other groups.

Protons and neutrons in an atomic nucleus are not simply squashed together any old how, but are, like the electrons around the nucleus (see page 141), arranged into 'shells' with specific capacities. A filled shell of protons or neutrons gives the nucleus a higher-than-average binding energy, so those nuclei are less liable to decay. These filled shells occur for proton and neutron numbers of 2, 8, 20, 28, 50 and 82. For protons, 114 was also predicted in the early 1970s to be a 'magic' number, as is a neutron count of 184. So the isotope ^{298}114 should be 'doubly magic' and especially stable; some early estimates put its likely half-life at several years, implying that, if it can be made, large quantities of this substance might be accumulated. The element-makers have not yet reached the 'island of stability' that supposedly surrounds this isotope of element 114; but it now seems that, even if it exists, the half-lives might not be quite as dramatic as some predictions imply. It has even been suggested that 114 might not be a magic number at all, and that the next closed shell of protons does not occur until 126. All the same, the isotope of element 114 created in 1999 appears to have a half-life of about 30 seconds – much longer than the fleeting lifetimes initially seen for elements 107–112.*

Fast work

What can we find out about the chemistry of these new superheavy elements? The isolation of mendelevium in 1955 was a chemical *tour de force* that, by including a chemical separation process, revealed something about this element's chemical behaviour. But in progressing from mendelevium (element 101) to nobelium (102), there was drop in half-life from tens of minutes to a few seconds. Subsequent research on nobelium, lawrencium and rutherfordium has uncovered isotopes with half-lives of over a minute, and in one case (^{262}Lr) over three hours. But, initially at least, the element-makers are confronted with very ephemeral materials. And the longest-lived isotope of rutherfordium known so far, ^{251}Rf, has a half-life of just 78 seconds. Can any real chemistry be conducted on timescales like this?

To make matters worse, the fusion events between the beam of high-energy ions and the target used to make these elements are very rare. Most of the ions simply miss the nuclei altogether. Even when

* Lifetimes for element 112 are still being debated. According to the most recently reported data, the three longest-lived isotopes of element 112 have half-lives of between 4 seconds and 11 minutes.

they do hit and fuse, the resulting nucleus is 'hot' and tends to decay by fission; only a tiny fraction of the 'hits' survive the cooling process. This means that it takes a lot of bombardment even to make a single atom of the new element. For rutherfordium and dubnium, typical production rates at the GSI installation are a few atoms per minute; for seaborgium, this falls to a few atoms per day. And these precious atoms are formed alongside a whole lot of 'junk' created by other nuclear reactions. To conduct any kind of chemical analysis of the artificial element, it must be separated from the junk.

The best way to speed everything up is to make it automated. This approach to rapid radiochemistry began in the late 1960s, both in the American and in the Soviet laboratories. There are four basic steps in any procedure of this sort:

1. Synthesize the element by ion-beam collisions with a target.
2. Transport the element, or its compounds, rapidly to the analytical chemistry apparatus.
3. Isolate and purify the element as quickly as possible.
4. Detect the element by monitoring its radioactive decay.

At Dubna, Ivo Zvara and his colleagues created an apparatus for performing chromatographic separations on gaseous compounds of short-lived elements, and they used it to study the chemistry of rutherfordium. Everything happened in a long tube, down which the Rf atoms were blown by an inert 'carrier' gas of nitrogen. At one end of the tube was the ion-beam target: a foil containing plutonium, which was transformed into rutherfordium by the impact of neon ions. Two gaseous chlorine compounds – niobium and zirconium chloride – were added to the target chamber, providing a source of chlorine that reacted with rutherfordium to make a volatile compound that was then carried by the gas flow down the tube, though a filter, and into a chamber containing sheets of mica. As rutherfordium decayed by alpha emission, the energetic particles drove into the mica to create tiny channels, which were later enlarged by etching so that they could be seen under a microscope. Fusion reactions in the target produced a rutherfordium isotope with a half-life of just two seconds or so* – but within a shorter time than that, these atoms could be carried from the target into the detection chamber.

* The Soviet team initially claimed that their Rf had a half-life of only 0.3 seconds, but later analysis cast doubt on this.

In 1970, Ghiorso and his co-workers discovered an isotope of rutherfordium (^{261}Rf) with a lifetime of about 70 seconds. This was just about long enough to survive liquid chromatography using an ion-exchange column. Ghiorso collaborated with Robert Silva at the Lawrence Radiation Laboratory in Berkeley (later renamed the Lawrence Berkeley Laboratory) to perform ion-exchange chromatography on about 100 atoms of ^{261}Rf. To get the Rf atoms as quickly as possible from the ion-beam chamber to the column, they fixed the target to a cylinder that could be driven pneumatically along a pipe, like a rabbit running through its burrow. At the far end of the pipe, a chromatography solution flowed over the 'rabbit', dissolving the Rf and carrying it to the column. Here, a resin grabbed tightly to metal ions with a charge of +4 or more (that is, ions with a valency of four). The column held onto rutherfordium while the other elements formed in the nuclear bombardment, such as californium and curium, were flushed away. This showed that, as predicted, Rf was not like the actinides, which tend to form ions with a charge of +2 or +3. Rutherfordium is the first *transactinide* element, and it more closely resembles the elements that lie above it in the fourth column of the periodic table: titanium, zirconium and hafnium, which also have a valency of four (Figure 9).

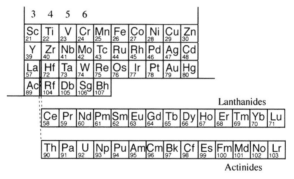

Figure 9 *Rutherfordium is the first element after the actinide series, and its position in the periodic table suggests that it should be similar to the group 4 metals, particularly hafnium (Hf). Likewise, dubnium was expected to resemble tantalum (Ta) in group 5. Some of these expectations were borne out by fast chemical analysis; but some were not. Dubnium in particular seems to deviate from the trends embodied in the periodic table – its chemistry may sometimes resemble that of niobium, and sometimes that of the actinide protactinium (Pa). Is this a sign that the structure of the table breaks down for these superheavy elements? That was the question underlying the study of seaborgium's chemical properties: would it too prove anomalous, or would it mimic the group 6 elements molybdenum (Mo) and tungsten (W)?*

Likewise, in 1987 a team at the Lawrence Berkeley Laboratory led by Ken Gregorich showed that dubnium-262, with a half-life of 34 seconds, behaves like tantalum (Ta) and niobium (Nb) in the fifth column (group 5) of the table (Figure 9). Although the Berkeley team conducted their experiments by hand, it took them just under a minute between ending the accumulation of Db in the target and detecting its alpha decay after a chemical reaction.

Bucking the trend

The question of how the transactinide elements fit into the periodic table is not quite so simple, however. In collaboration with Gregorich's team, researchers in Europe went on to make a detailed comparison of the chemical behaviour of dubnium, which they compared not only to Ta and Nb but also to the actinide protactinium (Pa). Because it sits directly underneath Ta, Db might be expected to resemble this element more than it does Nb. But in a sense Db also sits 'beneath' Pa in the actinide series (Figure 9). Is it truly a transactinide element that behaves as though it is in group 5, or does it still show any lingering resemblance to the actinides? In some of the chemical studies, which looked at the behaviour of complex ions formed by the combination of these metals with fluoride, Db seemed to resemble Nb more closely than Ta. In certain situations, it even showed some chemical kinship with Pa – that's to say, in such cases it doesn't seem to be a column 5 element at all, but appears rather to act as though it belongs in an extension of the actinide series.

This is very odd. As I have indicated, the periodic table is a map of the trends in chemical behaviour of the elements. Dubnium seemed to be defying those trends – it wasn't so clear where, on the map, it belonged.

What could be causing this anomalous behaviour? As we shall see in Chapter 8, the chemical properties of an atom are determined by the energies and arrangement of its outermost electrons, which are the ones involved in forming chemical bonds or ions. A trend in chemical reactivity like that seen for the alkali metals (group 1), which get increasingly reactive down the column, can be explained by the fact that, as the atoms get bigger, the lone electron in their outermost shell is ever less strongly bound, being progressively further from the nucleus, and so is easier to remove to form a +1 ion.

But for very heavy elements, a new and remarkable factor comes into play. The nuclei of these atoms have an immense amount of positive charge: there are 105 protons in the nuclei of dubnium, compared to just a few protons for elements such as carbon and nitrogen. This high charge means that electrostatic forces bind the electrons in the *innermost* shells very tightly indeed. Rather as a pirouetting ice skater spins faster the more tightly she contracts her body, so these innermost electrons in heavy atoms acquire very high velocities as they 'orbit' the nucleus in their so-called orbitals. These velocities can become so great that the electrons start to feel the effects of relativity.

According to Einstein's theory of special relativity, an object travelling at a significant fraction of the speed of light starts to gain mass. Inner electrons in very heavy atoms move fast enough to experience this effect – the average velocity of the electrons in the smallest orbital of a uranium atom is about two-thirds the speed of light – and the consequent increase in mass causes them to be drawn even closer to the nucleus. Thus, because of these relativistic effects, the inner orbitals contract.

That turns out to have an opposite influence on the outer electrons. These electrons don't feel the full electrostatic tug of the highly charged nucleus, because there are intervening shells of negatively charged electrons to partly counterbalance the attraction. In other words, the inner electrons 'screen' the outer ones. Contraction of the inner orbitals because of relativity has the effect of enclosing the nucleus in a tighter 'blanket' of screening charge, and so the nucleus's grasp on the outer electrons is weakened. So these orbitals, in turn, expand slightly: the outer electrons are less strongly bound. That modifies the chemical behaviour.* Thus, as relativistic effects kick in around the lower reaches of the periodic table, trends in chemical behaviour can be disrupted.

This seems to be what happens for dubnium. There is also some indication that rutherfordium may be affected by relativistic effects: the chloride compound of Rf ($RfCl_4$) is more volatile (that is, it evaporates at a lower temperature) than that of hafnium, whereas the straightforward trend from the periodic table predicts that the Rf compound should be *less* volatile. These anomalies mean that, in effect, the periodic table itself seemingly begins to break down for transuranic elements: the positions of Rf and Db in the table fail to be a wholly reliable guide to their chemical properties. Do such effects

---

* It can modify an element's physical properties too. Relativistic effects in gold (element 79) are responsible for its yellowish tint – without this, gold would have the whiteness of silver.

persist, and perhaps even increase, for the other superheavy, artificial elements?

Searching for seaborgium

All of this previous work on the human-made elements means that the experiments on seaborgium that I want to celebrate here are very much part of an evolutionary process in the rapid chemical analysis of incredibly tiny amounts of material – they do not, in a sense, represent a revolutionary advance. But they demonstrate the breathtaking state of the art, providing chemical information about seaborgium isotopes with half-lives of just a few seconds, on the basis of detections of just *seven* atoms. In addition moreover, they have something important to tell us about perhaps the most fundamental issue in chemical science: the integrity of the periodic table.

For more than 20 years since its discovery, the longest-lived isotope known for seaborgium was ^{263}Sg, whose half-life of about 0.9 seconds seemed to preclude any attempt at experimental studies. But, in 1994, scientists at Dubna and the Lawrence Livermore National Laboratory figured out how to make ^{265}Sg and ^{266}Sg, which have somewhat more amenable half-lives of 7.4 s and about 10–30 s respectively. For the chemical experiments at GSI, which began in late 1995, these isotopes were made by firing neon-22 ions into a target of curium-248. This project involved researchers from GSI, headed by Matthias Schädel, along with others from the Dubna and Berkeley groups, and from Germany and Switzerland (Figure 10).

We have seen that deducing the chemical behaviour of these ephemeral elements is all about making comparisons with the properties of lighter, stable elements. If the periodic table is any guide (that is, if relativistic effects do not wreak havoc), seaborgium should resemble molybdenum (Mo) and tungsten (W) more than it does the actinides (Figure 9). That is to say, while the latter tend to form ions with charges of +2 and +3 (divalent and trivalent), Sg should have a valency of six. Now, Mo and W react with chlorine and oxygen gas to form neutral oxychloride molecules such as MoO_2Cl_2 and WO_2Cl_2. These compounds are volatile – they become gaseous at just 150–200°C. So if seaborgium behaves in the same way, it will form comparable volatile compounds.

Rapid analysis of the 'gas-phase' chemistry of artificial elements is the speciality of Heinz Gäggeler of the Paul Scherrer Institute in

Figure 10 *Researchers at GSI in Darmstadt pose with the equipment they use to investigate the chemistry of artificial, superheavy elements. The ion-beam apparatus used to irradiate the curium sample and generate seaborgium lies just beyond the top of this image. In the upper right is the gas-flow system used to transport seaborgium-laden aerosol particles to the gas-chromatography apparatus, housed behind the white panel in the centre of the picture*
(Photo: Gabriele Otto, GSI.)

Villigen, Switzerland, who has developed an automated system called the On-Line Gas Chromatography Apparatus (OLGA). In collaboration with the GSI team, Gäggeler and his co-workers used various incarnations of OLGA in the 1990s to study the chemistry of ruther-fordium and dubnium; now they used OLGA Mark III to look at seaborgium.

Like ion-exchange liquid chromatography, gas chomatography separates elements according to how long their compounds take to pass through a column, to which the compounds may stick with varying tenacity. Each compound has a distinctive time of passage through the column, called the retention time. Thus, any volatile compounds of other elements formed in the ion-beam collisions would have a different retention time from seaborgium compounds and would exit the column at a different time. If, on the other hand, Sg did not form a volatile oxychloride at all – if it proved to be *unlike* Mo and W – then no Sg would show up at the far end of the column, where detectors were mounted in OLGA III to look for alpha particles with the energies characteristic of the decay of ^{265}Sg and ^{266}Sg.

The researchers calculated that the GSI ion beam would produce about one atom of ^{265}Sg or ^{266}Sg per hour. So they would have to detect each atom one at a time, and precious few in total. The collisions knocked these atoms clean out of the target foil, and they were then collected on tiny particles of carbon (aerosol) carried along in a flow of helium gas. The aerosol particles travelled down the tube to an inlet valve that admitted a mixture of chlorine, thionyl chloride and oxygen gases. The researchers anticipated that these substances would react with Sg atoms on the aerosol particles to form oxy-chlorides, and the particles were then carried by helium into a tube heated to 1000°C. That should be hot enough to evaporate any oxychloride compounds from the aerosol, enabling them to pass into the chromatography column (Figure 11). The whole process so far took about three seconds from the moment Sg atoms were formed in the target foil.

Only a few seconds later, the putative oxychloride compounds would leave the far end of the column and be deposited again on aerosol particles, which would carry them onto a thin plastic film mounted on a rotating wheel that presented each sample to the alpha-particle detectors. After several days of testing and running the OLGA III experiment, the researchers at GSI notched up four detection events: four atoms of Sg passed all the way from the target to the detectors. Three of these events showed the alpha-particle signature of ^{265}Sg decay; the fourth corresponded to the decay of ^{266}Sg. Thus seaborgium, like molybdenum and tungsten, did indeed form at least one kind of volatile oxychloride compound, which the researchers presumed (but could not prove) to be SgO_2Cl_2.

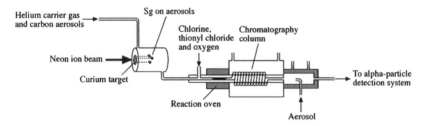

Figure 11 *The gas-chromatography apparatus OLGA III used to study the chemical properties of seaborgium*

OLGA was not the only instrument capable of probing the chemistry of seaborgium, however. In the late 1980s, Schädel and colleagues developed a system called the Automated Rapid Chemistry Apparatus (ARCA) for studying the chemistry of short-lived elements in solution. Like Silva's device in the 1970s, it enables man-made atoms to be passed quickly from their point of formation in ion-beam collisions to a solution that drips through an ion-exchange column, with alpha-particle detectors at the far end to detect radioactive decay events. In the computer-controlled ARCA II system used for the seaborgium experiments, the ion-exchange columns were just 8 milli-metres long, so that atoms reached the detectors just 45–90 seconds after their formation. Even this was too long for seaborgium atoms to survive the journey; but their initial presence could be inferred from the characteristic alpha signature of the first decay products of ^{265}Sg:^{261}Rf followed by ^{257}No. If the decay of these atoms were spotted, it meant that ^{265}Sg had made it through the column.

Schädel and colleagues carried out ion-exchange chromatography on solutions of the nuclear-reaction products in dilute nitric and hydrofluoric acids. They filled ARCA's columns with resins that grasp onto positively charged ions but not negatively charged ones or neutral compounds. The columns would retain the 2+ and 3+ ions of actinides, and so if Sg behaved like one of these elements it too would stick to the resin. So would uranium, which dissolves in a mixture like this to form the positive ion UO_2^{2+}. That was important. Since dubnium evidently behaves sometimes like its analogue in the actinide series (protactinium), then similar behaviour from seaborgium would make it act like uranium (Figure 9), perhaps forming SgO_2^{2+}. If, however, Sg was chemically similar to tungsten and molybdenum, then in this solution of acids it would form negatively charged complex ions: SgO_4^{2-} or oxyfluoride ions such as SgO_3F^- or $SgO_2F_3^-$. These would pass straight through the ion-exchange column.

And that is just what happened: in more that 5000 experimental runs, the researchers detected three distinct decay events correspond-ing to ^{261}Rf and ^{257}No in the liquid collected from ARCA II's columns. Three atoms of Sg had made their way through the columns, only to decay before they reached the detectors.

Thus, from a total of just seven Sg atoms identified by OLGA III and ARCA II, the researchers concluded that seaborgium, element 106, is chemically comparable to the group 6 elements molybdenum and tungsten. Surprisingly, there was no sign that relativistic effects, which modify the properties of dubnium, have the same influence in seaborgium. It was as though the periodic table had returned to normal – its architecture, said Schädel, was intact.

Into the unknown

But does it stay that way? What happens after seaborgium? In 2000, the teams from GSI, Berkeley, Villigen and Dubna gathered forces again to address the next challenge: element 107, bohrium. Would it be comparable to the elements above it in column 7 of the table, technetium and rhenium? Until that year, the longest-lived isotope known, ^{264}Bh, had a half-life of less than half a second, so that even the swiftest apparatus would be to no avail. But then ^{267}Bh was discovered at the Lawrence Berkeley Laboratory, formed by colliding ions of neon-22 with a target enriched in berkelium-249. It has a half-life of about 17 seconds. Gäggeler and his colleagues instantly took up the challenge of studying its chemistry, and they used OLGA to show that bohrium behaves like a typical group-7 element, forming a volatile oxychloride BhO_3Cl by reacting with hydrogen chloride and oxygen. Because the formation rate of ^{267}Bh was so slow, the experiments took a month to conduct, and even then the researchers had to rely on detections of just six atoms.

So once again the periodic table proved to be on firm footing. Two years later the same teams caught a glimpse of the chemistry of element 108, hassium (Hs), for which the isotopes ^{269}Hs and ^{270}Hs have half-lives of 10 and 4 seconds respectively. Seven detection events served to show that hassium, like osmium above it, forms a volatile tetroxide compound, HsO_4.

This journey into the 'artificial' periodic table shows no signs of approaching an end. Even though the initial discovery of new human-made elements generally offers nothing but the briefest glimpse of a transient isotope, it has so far proved to be the case that longer-lived isotopes will subsequently come to light. Indeed, elements 104 to 108 are predicted by theory to have some isotopes, rich in neutrons, with half-lives measured in days or even years. The discovery of such atoms (provided that their production rates aren't prohibitively slow) could enable a more leisurely and thorough perusal of their chemical behaviour. Moreover, if there is anything like an island of stability around element 114, chemical studies on these still heavier elements might become even easier. So long as isotopes with half-lives of several seconds can be found, apparatus like OLGA and ARCA will be able to tell us something about their chemistry. Experiments currently being conducted at GSI may soon tell us whether or not element 112 acts like mercury, the element above it.

In some ways this is chemistry as we have always known it: the investigation of how atoms combine. But in another sense it is something quite new, for these are atoms that nature cannot make. It would have astonished Lavoisier, who helped to define the modern notion of an element, to think that future scientists could create elements of their own – just as it would have amazed his near-contemporary John Dalton to think that we would be able to pick up his atoms one at a time and watch what they do. It would have only enhanced their wonder that these feats are made possible by a form of alchemy, the transmutation of one element into another.

There is, however, another novel aspect of the technical tour-de-force that enables the investigation of seaborgium, bohrium and their ilk that does not depend at all on their artificial nature. One can regard these fantastic experiments as examples of the tendency for today's chemistry to probe *extremes of scale*. Modern instruments allow researchers to see individual molecules and atoms, to move them, to investigate their behaviour. We can watch – literally, we can watch the movie – as two molecules come together and react. It is possible to perform chemical surgery on single molecules: to cut them up or to stick on new chemical groups. We can see a single fluorescent molecule blink on and off as it absorbs and emits photons of light. We can measure individual chemical bonds as they form and break, and measure their strength just like measuring the strength of a piece of string. We can study the movements of these molecules on time-scales appropriate to such tiny objects, following a chemical bond during the single in-and-out motion of a vibration that happens more than a trillion times a second.

These are things that nineteenth-century chemists could not even conceive of, for they hesitated to allow that atoms and molecules are real entities, as opposed to convenient abstractions that one could represent with wooden balls and sticks. Even as the first human-made elements were being created, scientists remained insistent that the study of matter was a statistical science whose law-like regularities depended on our viewing huge numbers of particles at once. In a sense, experiments that involve just half a dozen or so radioactive atoms confirm this notion, for the lifetimes of the seven seaborgium atoms in the GSI experiment are quite different from one another: one of them decayed after just 0.6 seconds, another after 27 seconds. A half-life says nothing about when an individual atom will decay, but only when, on average, we might expect it. So scientists studying matter one atom or molecule at a time are indeed conducting a new kind of chemistry. Like social scientists switching from population censuses to personal, face-to-face interviews, they are forced to confront the idiosyncrasies of individuals, who sometimes behave in ways we can never quite predict.

The Chemical Theatre

I discovered that chemistry was dramatic when I tried to make sodium chloride during my school lunchtime. I plunged a naked lump of sodium metal into a gas jar of green chlorine. There was a crack like thunder, rattling the glass of the gas cupboard (I'd at least been trained to exercise that much caution), and I found that the gas jar was neatly sheared off at its base.

It could have been worse. Humphry Davy, who discovered that chlorine was an element in 1810, was fascinated to hear from his friend André-Marie Ampère that it will form an explosive compound with nitrogen. Undaunted by the fact that the man who discovered this, Pierre Louis Dulong, lost an eye and a finger in the process, Davy set about investigating this substance with his new assistant, Michael Faraday. As James Hamilton describes it in his biography of Faraday, Davy told his young assistant to combine the explosive nitrogen trichloride with ammonia:

> This immediately produced thick acrid smoke, ammonium chloride, which filled the laboratory, making them both choke violently. Once the smoke had cleared they took some more glass bowls and tubes and tried again with ammonia. The smaller tubes constrained the reaction, but in an instant the whole lot exploded .... The next day they tried again, this time with yet more violent results. There were four big explosions in the laboratory that day, audible throughout the building ... the day's work came to an abrupt end when Faraday had his hand nearly blown apart.

But bangs are what people expect from chemistry; no 'demonstration lecture' is complete without a few of them, preferably accompanied by a liberal excess of light and smoke. Leonard Ford's classic 'manual for chemical magicians', first published in 1959, contains prescriptions for several explosions, including various firework recipes and a somewhat safer version of Davy and Faraday's experiment that uses the triiodide rather than the trichloride of nitrogen. (I tried this in my youth, naturally; I did not exactly blow my hand apart with it, but was left with stinging fingers, purpled by iodine.)

Ford's book drew on his experience of staging 'travelling shows for science', in which he would bedazzle crowds at service clubs, youth organizations and churches with his chemical magic. (One wonders how a man who could spit fire would have been greeted in the American South in the 1950s.) For Ford, chemistry was theatre.

Or rather, you might say, it was entertainment with a pedagogical message, which is not exactly the same thing. Yet science has truly found its way into the theatre today. Tom Stoppard drew metaphors from quantum physics in his play *Hapgood* (1988) and from chaos theory in *Arcadia* (1993), while Michael Frayn's *Copenhagen* (1998) played with the notion of uncertainty in dramatizing the meeting between Niels Bohr and Werner Heisenberg during the Second World War. A character who muses on the complexities of cosmology or number theory seemed for a time almost to have become *de rigeur* for middlebrow productions in London's West End.

Could this use of science in theatre be an older tradition than is generally appreciated? Christopher Marlowe's *Dr Faustus* (published in 1604) and Ben Jonson's *The Alchemist* (1610) are commonly cited as early examples of the genre, but Marlowe was retelling a folk legend, while Jonson, however well versed in alchemy himself, was merely using its popular image of charlatanry as a vehicle for his boisterous farce. More pertinent to the contemporary tradition of using scientific metaphors in drama is Charles Nicholl's claim that the alchemical allegories of the late Renaissance inform some of Shakespeare's famous works. The 'transformation of the king' is a recurring theme in alchemy: the processes taking place in the alchemist's retorts and crucibles were typically described in allegorical terms as a series of travails and indignities suffered by the 'king', representing alchemical gold (Figure 12). 'The whole unfolding process of King Lear', Nicholl claims, 'is deeply and intentionally alchemical'. Certainly, Shakespeare knew about alchemy: he explores alchemical themes in his sonnets, while Friar Laurence in *Romeo and Juliet* is the archetypal Paracelsian chymist who prepares medical remedies and potions by distilling herbs. Shakespeare's sometime patron, Sir George

Figure 12   *The trials and tribulations of the 'king': a set of allegorical engravings from Michael Maier's* Atalanta fugiens *(1618), illustrating the chemical processes of alchemy*

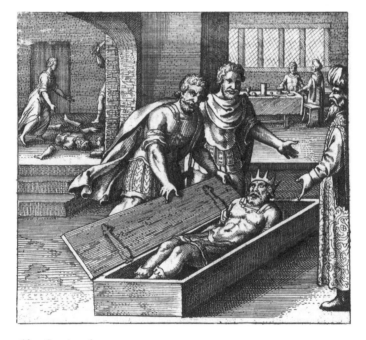

Figure 12 *Continued*

Carey, was an enthusiast of the alchemical ('Hermetic') arts, and Carey's friend, the Elizabethan alchemist Simon Forman, was alleged to have provided the model for the mountebank Subtle in Jonson's satirical depiction of alchemy.

Even if Nicholl's argument holds, it might be stretching a point to say that this is truly a forerunner of 'science in theatre'.* But there is a deeper connection between the task of the dramatist and the 'theatrical' element of experimental science. The classic advice to the playwright is 'show, don't tell': the emotional and psychological truth of the drama must emerge through the characters' actions, not via explicatory narration or dialogue. We can rationalize what we have seen after the event, but we must first experience it directly. Similarly, an alchemical experiment was, for Renaissance intellectuals, a source of direct revelation. In contrast to the mechanical philosophy of René Descartes and his followers in seventeenth-century Paris, alchemy enabled the skilled adept to leapfrog plodding reason and gain a direct glimpse of the true order of the universe. This was a form of gnosis, akin to (and indeed allied to) mystical religious movements in which such epiphanies were held to be the only way humankind could come close to appreciating God's wisdom and design. Where the religious mystic saw God, the alchemist saw nature, unveiled as it were in a blinding flash. And that kind of insightful vision is what a good experiment can still provide today.

This does not mean, of course, that experimental science need involve short-circuiting our rationality. Yet all the same, the immediacy of a well-conceived experimental demonstration connects with our innate ability to comprehend by visual revelation: seeing is 'believing'. There is surely a gnostic element to the power of an experiment like that of Ernest Rutherford, described in Chapter 4, where the result is heralded by the sudden appearance of helium's spectral glow. Even the seemingly prosaic movement of an automated graph-plotting pen as it sweeps across the page to trace out the peak you were looking for (or maybe one you were not) excites in the experimenter a feeling of illumination and wonder. It is more than the excitement at having your hunch confirmed; it is as though (indeed, it literally *is* that) the universe has just revealed a new part of itself.

---

* This term has been coined by chemist Carl Djerassi to describe the various plays that he has written or co-written on scientific themes, notably *Oxygen*, which Djerassi and Nobel laureate Roald Hoffmann wrote in 2001.

That sense of 'illumination', echoing the Neoplatonic reverence of light and the belief of Paracelsian proto-chemists in the 'Light of Nature', informs the 'scientific' paintings of the artist Joseph Wright of Derby. The expression on the face of Hennig Brandt as he discovers phosphorus in Wright's *The Alchemist in Search of the Philosopher's Stone* (1771) (Figure 13) conveys religious awe: white-bearded and kneeling in his robes, he is an Old Testament prophet seeing the burning bush, not a scientist making a discovery. Even though Wright is ostensibly portraying the isolation of this new element in the 1670s, his title suggests that the picture is really all about the Promethean quest for divine, revelatory knowledge.

The same quasi-religious atmosphere pervades Wright's *Experiment on a Bird in the Air Pump* * (1768), even though this painting shows a more recent episode in science history. The scientist here has the flowing, silver hair of a mystical savant, and the scene shows a ritualistic sacrifice of a pet bird from which a young girl averts her eyes. The dramatic lighting is even more gratuitous than it is in Wright's depiction of Brandt's laboratory, which makes the scene all the more explicitly gnostic.

This powerfully theatrical image – the light of revelation, as in the radium effulgence that lit up Marie Curie's chilly shed – is still with us. Here is how English playwright Stephen Poliakoff (brother of chemist Martyn Poliakoff) shows his audience the 'Sun Battery', which uses sunlight to convert water into hydrogen, in his 1996 play *Blinded by the Sun*:

> Muzak. The equipment is brought on, supervised by Barbara. A simple wooden table, completely bare, except for a metal stand and clamp holding a glass tube. The muzak really wells up. It is coming from elsewhere in the building, a radio blaring out light orchestral music. A beam of light stabs across the darkened stage. The tube starts bubbling.

The theme of illumination in these images of experimental chemistry is accompanied by another, equally profound and emotive theatrical device: the ritual. The theatre, says director Peter Brook, cannot survive as a vital art without a sense of ritual, 'the notion that the stage is a place where the invisible can appear'. Alchemy was

---

* http://www.nationalgallery.org.uk/cgi-bin/WebObjects.dll/CollectionPublisher.woe/wa/largeImage?workNumber=NG725&collectionPublisherSection=work

Figure 13   *The Alchemist in Search of the Philosopher's Stone (1771), by Joseph Wright of Derby*
(Reproduced with permission from the Derby Museum and Art Gallery)

undoubtedly ritualistic: there was a proper order to the experiments involved in transmutation, and one could not expect success if this sequence was violated, regardless of one's skill. In Renaissance science, still blended with the discipline of natural magic, there was a sense that experimentation involved a kind of propitiation of nature, in much the same way that a church ceremony had to observe the correct order of service if it was to please God. This affinity of the early scientist with the priest was made explicit in Francis Bacon's *New Atlantis*, in which Bacon presented a vision of a scientific brotherhood that was possibly influenced by occult movements such as Rosicrucianism. On Bacon's 'Atlantis', an island called Bensalem in the Pacific Ocean, the utopian society is, as we saw earlier, governed by the theocratic scientist-scholars of Salomon's House, who make marvellous devices and inventions by means of their experimental 'art' while observing a strict code of secrecy to prevent their knowledge from falling into the wrong hands. And it is surely not mere whimsy that impels the Royal Institution in London, where Davy and Faraday perfected the art of scientific lecture as a form of theatre, to maintain its ritualistic tradition in the Friday Evening Discourses. A Discourse begins with the grand entry of the speaker in full evening dress as the clock chimes eight, and the speaker is supposed to conclude just as the clock strikes an hour later.

Ritual links us to the past: history gives an action a form of confirmation. The experimental scientist typically undergoes an apprenticeship that is only partly about acquiring useful skills; it is also concerned with the inheritance of a tradition. My final-year university research project in chemistry was almost entirely theoretical, and yet I was expected, like everyone else, to first take a course in glass-blowing (which in consequence I now appreciate as an art). The Italian writer and chemist Primo Levi was aware that, by conducting the same processes and manipulations of matter that his chymical and alchemical predecessors devised hundreds of years before, he was enacting a time-honoured ritual that somehow validated the business of the laboratory:

> Distillation is beautiful. First of all, because it is a slow, philosophic, and silent occupation, which keeps you busy but gives you time to think of other things, somewhat like riding a bike. Then, because it involves a metamorphosis from liquid to vapour (invisible), and from this once again to liquid; but in this double journey, up and down, purity is attained, an ambiguous and fascinating condition, which starts with chemistry and goes very far. And finally, when you set about distilling, you acquire the consciousness of repeating a ritual consecrated by the centuries.

And there seems to be some ritualistic collective memory that has kept alive the most ancient of the images conventionally used to depict

Figure 14 *The image of the 'gazed-at' flask is the standard way of showing a chemist at work. The pose derives from that used in illustrations of medieval doctors, as here (top left)*
(Reproduced with permission from the National Library of Medicine (top left) and CORBIS (top right))

the chemist at work – the scientist gazing into a raised beaker of fluid (Figure 14). This posture, repeated in countless stock photographs, is very evidently derived from the standard portrayal of the medieval physician (Figure 14, top left), who made his diagnoses by inspecting a flask of the patient's urine. The physician became the chemist by way of 'iatrochemistry', the medical chemistry practised in the European colleges of the early Enlightenment (and where, while we're about it, do you think *that* term came from?). In the Chemical Theatre, old habits die hard.

CHAPTER 6

Molecules Take Shape

Pasteur's Crystals and the Beauty of Simplicity

Paris, 1848—A young French scientist named Louis Pasteur is investigating the shapes of crystals. He is puzzled as to why different substances form crystals of the same shape, while at the same time some identical substances form crystals of different shapes. He suspects these properties are related to the shapes of the constituent molecules. His studies of crystalline salts prepared from the by-products of wine-making thus lead Pasteur to a crucial insight about the three-dimensional structures of carbon-based molecules.

When, as I mentioned in the introduction, The American Chemical Society, through their bulletin *Chemical and Engineering News*, recently polled readers and experts to identify the most beautiful experiment in the history of chemistry, they responded by giving the highest ranking to Louis Pasteur's separation of mirror-image molecular forms of tartaric acid, conducted just a year after Pasteur gained his doctorate at the École Normale in Paris. The experiment apparently lays claim to a special beauty because it was conceptually simple yet painstakingly executed, and because it revealed a deep and important truth in chemistry in a way that was clear-cut and unambiguous.

Because all of these things are essentially true, I believe that Pasteur's experiment earns its place in this book. But I am fairly sure that the choice of many of *C&EN*'s respondents was based on the textbook account of the event, in which Pasteur's insight is a stroke of genius that arrives as a eureka-like revelation. It may come as a disappointment, then, to learn that things did not, in all probability, happen that way.

The reality was more like science as it is usually conducted: a rather mundane and clumsy affair in which the significance (or otherwise) of one's results becomes apparent only in retrospect. It is understandable that the history books should be misleading, however, because they quite reasonably assumed that they could rely on the first-hand account of the experiment given retrospectively by Pasteur himself – a story that was subsequently embellished, with Pasteur's blessing, by his son-in-law in the first of several hagiographies of this towering figure of French science. We'd do well to heed the advice of science historian Gerald Geison, who has carried out the detective work that reveals the real story of Pasteur's first great discovery: 'to be ever skeptical of participants' retrospective accounts of scientific discovery'. While this was by no means the only case of self-mythologizing in nineteenth-century science, few people were more adept at arranging facts to suit their own ends than Louis Pasteur.

Crystal gazing

There is no rule which says that great scientists have to be likeable, and for Pasteur (Figure 15) that is just as well. It is not that he was especially monstrous or wicked, but his arrogance and lack of charity do not create an endearing portrait. He was energetic and skilful at self-promotion, and somewhat calculating in the relationships he fostered (and in those he neglected). Only grudgingly did he give credit to others, but he was not slow to claim it for himself. Despite his tremendous self-assurance, he was easily stung by criticism. Yet he dished it out freely and sometimes vindictively, to the extent that one octogenarian French scientist was sufficiently enraged to challenge him to a duel (which was, fortunately, not enacted). Perhaps more damagingly for Pasteur's scientific reputation, he was willing to rewrite history and was not averse to a rather biased selection of the evidence to support his ideas. But, despite all that, his genius is undeniable. It is after all commonly the case that genius must be accompanied by a degree of assertiveness if it is to be recognized at all.

Although remembered by posterity primarily for his biological discoveries – he more or less founded the field of microbiology, and made pioneering contributions to immunology and the technique of vaccination – and although lauded by chemists, Pasteur actually began his career in physics. Like several later physical scientists who turned their attention to biology – Linus Pauling and Francis Crick come to

Figure 15 *Louis Pasteur, the very image of French bourgeois respectability*
(Reproduced Courtesy of the Library and Information Centre, Royal
Society of Chemistry)

mind – Pasteur's initial enthusiasm was for crystallography, the study
of crystals.

Born in Dôle in the region of Franche-Comté, Pasteur went to Paris
in 1844 to study for his doctorate at the prestigious École Normale.
One of his prime mentors there was Gabriel Delafosse, who had been a
student of the great crystallographer Abbé René Just Haüy. When
Haüy formulated his ideas about the shapes of crystals in the 1780s, it
was commonly believed that these reflect the shapes or arrangements
of their constituent particles – a notion that went back at least as far as
Johannes Kepler's speculations on the hexagonal symmetry of snowflakes in the early seventeenth century. Haüy believed that the form of
every crystal could be traced back to the shape of a fundamental
grouping of its particles, which he called an 'integrant molecule'. This
can be seen as analogous to the modern idea of a 'unit cell', the simplest

block of atoms or molecules that repeats again and again throughout the crystal like bricks in a wall.

In essence, Haüy's idea seemed to imply that the shape of a crystal depends on the molecules from which it is made. Chemists at the beginning of the nineteenth century had no clear idea of what molecules were, but John Dalton's atomic theory helped them to envisage compounds – substances that contained more than one element – as being composed of discrete groupings of atoms in fixed ratios. In 1818 the French chemist Michel Eugène Chevreul defined different chemical substances in terms of the kind, number and spatial arrangement of atoms in these molecules. But no one gave much thought to the spatial arrangement, for they knew of no way to investigate it; indeed, many chemists considered the molecular models of coloured wooden balls and sticks that became popular with chemistry lecturers as mere heuristic devices that should not be considered to bear any relation to the way actual molecules looked. They fretted that such 'toys' would be taken too literally by the uninformed. Even in 1869, William Crookes advised students to 'leave atoms and molecules alone for the present. Nobody knows how the atoms are arranged.'

Twisted logic

It had been known since the seventeenth century that some crystals have a curious effect on polarized light. When normal light is passed through a polarizing filter, the electromagnetic waves in the emergent rays all oscillate in the same plane, like so many snakes writhing in the horizontal plane of the ground. The Dutch scientist Christiaan Huygens, Isaac Newton's great rival in the study of light, found that when such polarized light is passed through the transparent mineral Iceland spar, the plane of polarization is rotated by a well-defined angle (Figure 16). This behaviour became known as optical activity. Quartz has the same effect, and it was widely assumed that optical activity somehow results from the spatial arrangement of the molecules in a crystal – they act rather like a spiral staircase that turns the polarization through a certain angle. Some materials will rotate the plane of polarization to the left, some to the right: there is a 'handedness', an asymmetry, to the effect.

In 1815 the French physicist Jean-Baptiste Biot found that some crystalline organic substances remain optically active when they are dissolved in solution. This apparently undermined the 'spiral-staircase' idea, showing in effect that the individual rungs of the staircase were

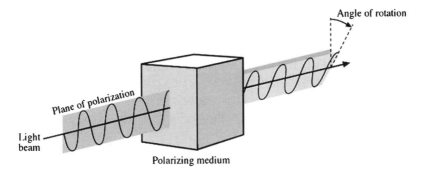

Figure 16 *'Optically active' crystals rotate the plane of polarized light – some rotate it to the left, some to the right*

themselves active: it wasn't a question of how they were assembled in the solid. Biot argued that the property of optical activity must be inherent in the organic molecules themselves, and he proposed that they had a twisted configuration of their atoms.

It has often been implied that Pasteur's doctoral thesis at the École Normale was an attempt to uncover the relation between optical activity and molecular 'twistedness'. But that wasn't so. Pasteur did indeed study this phenomenon in solutions of organic compounds, but the conclusion of his thesis was essentially a restatement of Biot's idea:

> I regard as extremely probable that the mysterious and unknown disposition of physical molecules, in a whole crystal of quartz, is also found in [optically] active bodies, but, this time, in each molecule in particular; that it is each molecule, taken separately in an active body, that must be compared, for the arrangement of its parts, with a complete crystal of quartz.

One year after he wrote these words, however, Pasteur had something much more momentous to add.

Under Delafosse's guidance, he investigated the shapes of crystals. Haüy's proposal that a crystal form is dictated by the nature of its constituent molecules was challenged by two discoveries made by the German scientist Eilhardt Mitscherlich in 1819–21. First, compounds composed of different atoms can show the same crystal shape – a property called isomorphism. Second, some compounds can crystallize in more than one crystal form, which was known as polymorphism. It rather looked, then, as though crystalline form was *independent* of chemical constitution.

In the 1840s, Delafosse was trying to revise his mentor Haüy's theory so that it could accommodate isomorphism and polymorphism, and he set Pasteur to work on the problem. One example of isomorphism that had particularly interested Mitscherlich was the case of the salts of the organic compounds tartaric and racemic acid. Tartaric acid salts, or tartrates, were long known as by-products of wine-making: they were precipitated as white solids on the walls of wine casks. Paracelsus in the sixteenth century named these precipitates 'tartar', and in the seventeenth century the French apothecary Pierre Seignette of La Rochelle identified one of these salts as sodium potassium tartrate.* Carl Wilhelm Scheele showed in 1770 that this 'Seignette salt' could be converted into an acidic substance, the first known organic acid, called tartaric acid. This mild acid became an industrially useful substance, used in textile manufacture and medicine. Wine makers began to collect and sell it as a profitable sideline.

That was how an industrialist named Kestner discovered racemic acid at his factory at Thann in Alsace. It appeared to be simply a second kind of tartaric acid, with the same elemental composition and almost identical properties, and Kestner gave some of it to the French chemist Joseph Louis Gay-Lussac to study. It was Gay-Lussac who named it after the Latin *racemus*: a bunch of grapes. But Jöns Jacob Berzelius, a doyen of chemical nomenclature, called this new compound paratartaric acid, and decided that it was an *isomer* of tartaric acid: that is to say, the molecules contained the same atoms, but differently arranged.

Although Mitscherlich decided that salts of tartaric and racemic acid were isomorphous, he was hard pushed to find any real difference at all between the two substances. Their crystals looked identical, and there seemed to be only one way of distinguishing them: tartaric acid (or a solution of a tartrate) was optically active while racemic acid was not. According to Biot's view, this must mean that molecules of tartaric acid were 'twisted' while those of racemic acid were not. Mitscherlich told Biot what he had found, and the Frenchman repeated his experiment and confirmed the result in 1844. Browsing in the library at the École Normale in 1848, Pasteur happened to see Biot's report, and he wondered how it could be that tartaric acid and racemic acid could give rise to salts with isomorphous crystals, identical in shape, if the molecules of the organic acids themselves had quite different shapes.

---

* 'Seignette salt' is also known after Seignette's locality as Rochelle salt, and it was one of the materials that Pierre and Jacques Curie found to be piezoelectric – see page 40.

Were the crystals really identical? That's what Mitscherlich had said. But Pasteur decided to check.

This is how Pasteur began his first great experiment. What was he really looking for, and what did he expect to find? In a retrospective lecture in 1860, Pasteur explained that he wanted to discover why it was that the two acids differed in their optical activity. But Geison has inspected his lab books, which initially make no mention of optical activity. Instead, Pasteur is caught up with a recondite question about how the isomorphism of salts of the two acids might be connected to the number of water molecules that get 'locked up' in their crystalline lattices. There is no grand hypothesis – his notes look like the typical records of a young researcher, rather dull collections of tables and measurements accompanied by marginal notes that signify a desperate quest to find some logical order to it all.

Pasteur used a microscope to inspect the shapes of crystals of sodium ammonium tartrate and 'paratartrate' (that is, the corresponding salt of racemic acid). He had learnt how to recognize and classify the facets of crystals, and he found that, just as Mitscherlich had said, these salt crystals were 'asymmetric': they had a handedness. If you start with a symmetrical crystal shape like a cube, and cut off some corners but not others, you can produce left- or right-handed shapes that are mirror images of one another (Figure 17a). Such crystal forms are said to be hemihedral. The tartrate salt crystals were always of the right-handed hemihedral form. But Pasteur recorded in his notebook that for the paratartrate salt, 'The crystals are frequently hemihedral to the left, frequently to the right' (Figure 17b). He added that sometimes the truncated corners 'repeat', so that left-handedness and right-handedness

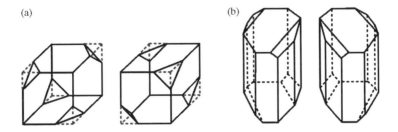

(a) (b)

Figure 17 *a, Hemihedral crystals have a particular handedness: the two forms here are mirror images, and cannot be superimposed on one another. b, The ideal mirror-image forms of crystals of sodium ammonium tartrate, the salt investigated by Pasteur in 1848*

cancel one another out to produce symmetrical crystals. But he later crossed out this comment.

It looks like a routine observation – puzzling, perhaps, but just a part of the wider puzzle. Compare this with the version of events that Pasteur later gave as the 'official' record. He asserted that, knowing that tartrate was optically active but paratartrate was not, he suspected that Mitscherlich had been wrong to say their crystals were identical. He thought that, on the contrary, the paratartrate crystals would prove to be symmetrical, with no handedness. When he looked and found that they were after all asymmetric, Pasteur claimed that 'for an instant my heart stopped beating'. But then he looked more closely and noticed that they were asymmetric in *both* senses: some were left-handed, and some right-handed, in contrast to the single handedness of tartrate. And so (Pasteur said), the answer to this riddle struck him in a flash. Right-handed tartrate revolved the plane of polarized light. Paratartrate, seemingly a mixture of left-handed *and* right-handed forms, left the light unchanged. Could this be because the effects of the two forms cancel each other out? That is to say, maybe the right-handed paratartrate crystals were composed of right-handed molecules, and the left-handed crystals of left-handed molecules, and the two were present in solutions of paratartrate in equal measure.

In that case, the molecules in the left-handed crystals would, if separated and dissolved, rotate polarized light one way, and the right-handed molecules would rotate it equally in the other direction. All one had to do was separate the two types of crystal. Pasteur did this by hand, wielding tweezers to patiently sort the crystals into two piles. Then he redissolved them – and there it was, the very result he'd anticipated.

In one account, he is said to have exclaimed 'Tout est trouvé!' – all is found. Other reports are yet more colourful: René Vallery-Radot, Pasteur's son-in-law, claims that Pasteur rushed from his lab, grabbed a hapless curator in the corridors of the École, and dragged him into the Luxembourg Gardens to tell him about the great discovery.

Pasteur was not a whimsical man; he maintained an air of stiff bourgeois respectability, sharpened by a rather imperious and abrupt manner. It is hard to imagine him, even in his youth, prancing about gaily with excitement. But Pasteur was happy for Vallery-Radot to tell history this way (he saw the proofs before the biography was published), and if it is hyperbole then it is not so very unusual for its time. Victorian scientists frequently indulged in romanticized retelling of their discoveries: they seemed to feel that a bit of mythologizing would do science (and them) no harm (see Divertissement 2).

We have to be comparably cautious about Pasteur's description of subsequent events. The head of his laboratory, Jerome Balard, conveyed the remarkable discovery to Biot, who asked to see the young man responsible for it. Brought into the presence of the great French scientist at the Collège de France, Pasteur was requested to repeat his experiment before Biot's eyes. He did so, whereupon (according to Pasteur) Biot, 'visibly moved, took me by the arm and said: "My dear boy, I have loved science so much all my life that this stirs my heart."' Yet even if the exact words were not so epigrammatic, the sentiment may have been much the same, for Biot subsequently regarded Pasteur as his protégé.

It is not in fact clear that Pasteur had given any thought to the relationship between optical activity and crystal shape in the tartrates and paratartrates until he performed his laborious separation of crystals – certainly, when he looked closely at their shapes, he does not appear to have formulated a strong hypothesis about whether paratartrate would have symmetric or asymmetric crystals. So we must doubt that his heart truly skipped a beat at what the microscope revealed. In addition, having prepared his two solutions of right- and left-handed paratartrate, his measurements of optical activity were not, so to speak, crystal-clear: the solutions did rotate polarized light in opposite senses, but not to identical amounts. Reasonably enough, Pasteur put this down to his limited ability to distinguish the two crystal types under the microscope*, but that cool comment doesn't exactly support the idea of a eureka-like epiphany.

Mirror matter

Be that as it may, Pasteur saw with perfect clarity what his observations implied. Paratartaric or racemic acid was not a single molecular substance, but two: it was a mixture of mirror-image molecules. The right-handed versions were identical to the molecules in tartaric acid; the left-handed molecules were in effect tartaric acid twisted

---

* Traditional accounts of the experiment (including an earlier one of my own) show the two mirror-image paratartrate crystals as the icons of geometric precision indicated in Figure 17b. In reality, the distinction is rarely this clear. Modern researchers have tried to repeat Pasteur's experiment, and have found that the crystals are much harder to tell apart than these diagrams suggest.

the other way, an isomer with equal but opposite optical activity. Pasteur introduced the word 'racemic' as a general term for an equal mixture of both isomers of an optically active compound.

He also understood what the notion of handedness must imply for a molecule's structure. One possibility was that the atoms were arranged in a screw-like pattern, a spiral staircase turning to the right or the left. This is indeed the atomic arrangement that makes quartz (silicon dioxide) optically active. But Pasteur realised that there were other options for organic molecules, as he explained in 1860:

> Are the atoms of the right-handed acid grouped along the turns of a right-handed helix, or situated at the vertices of an irregular tetra-hedron, or disposed according to such and such a fixed asymmetric [*dissymmetrique*] arrangement? We would not know how to answer these questions. But there can be no doubt that there is a grouping of the atoms of an asymmetric type that is not superposable on its mirror image.

In 1874 the Dutch scientist Jacobus van't Hoff did offer an answer. He suggested that the stick diagrams that chemists had begun to use to depict organic molecular structures as an arrangement of atoms in space should be extended into the third dimension, out of the flat printed page. Such drawings had evolved from a mere graphical representation of molecular categories – a way of showing how different molecules were related to one another – to the point where they were regarded (by some researchers, at least) as showing the actual shapes of molecules. The atoms were represented by chemical symbols – C for carbon, H for hydrogen, O for oxygen – and the links between them were shown as lines or dashes. It was recognized since the work of Friedrich August Kekulé in 1858 that carbon atoms generally form bonds to four other atoms. (The Scottish scientist Archibald Couper had the same idea independently, and would have published it first had his paper not been delayed by editors.) According to this scheme, tartaric acid would be depicted like this:

$$
\begin{array}{c}
\text{COOH} \\
| \\
\text{H}-\text{C}-\text{OH} \\
| \\
\text{HO}-\text{C}-\text{H} \\
| \\
\text{COOH}
\end{array}
$$

But van't Hoff proposed that, instead of having their four bonds arranged as a flat cross, carbon atoms disport their bonds pointing to the corners of a tetrahedron:

For carbon atoms that have four *different* atoms or groups of atoms attached (and this is true of the two central carbons in tartaric acid) there are two different ways to arrange the substituents in space, which are not superimposable. These are mirror images of one another:

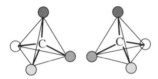

This, said van't Hoff, is how organic compounds can acquire a handedness that results in their optical activity. Just a month after he published this idea, Joseph Le Bel in Paris clarified what it implies: any molecule with an asymmetric carbon atom – one with four different groups attached – will be optically active *unless* there is another equivalent grouping in the molecule that cancels its effect. In the latter case, the molecule becomes symmetric, so that it is its own mirror image: it is no longer 'handed', and no longer rotates polarized light. Tartaric acid has such a molecular form, called the meso form, as well as the optically active right- and left-handed forms (Figure 18).

A molecule's 'stereochemistry' refers to the arrangement of its atoms in three-dimensional space: molecules containing the same atoms and the same connections between them, but different three-dimensional shapes, are known as *stereoisomers*. Two stereoisomers that are mirror images of one another are called *enantiomers* (from the Greek *enantios*, opposite). Lord Kelvin decided that the property of molecular handedness should itself be dignified with a classical name, and in 1904 he derived one from the Greek word for hand, *kheir*: this sort of asymmetry is called *chirality*. Thus the two optically active forms of tartaric acid are chiral molecules, and enantiomers.

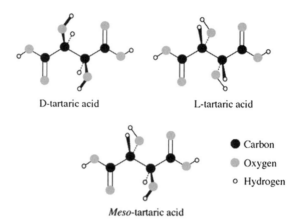

D-tartaric acid   L-tartaric acid

Meso-tartaric acid

● Carbon
◉ Oxygen
○ Hydrogen

Figure 18 *Optically active forms of tartaric acid (denoted D and L), and the optically inactive meso form. Here I have depicted the three-dimensional structure of the molecules by using thick black lines to denote bonds pointing up out of the plane of the page, and dashed lines to indicate bonds pointing down beneath this plane*

Pasteur's famous dictum that 'luck favours the prepared mind' is often quoted, but it is possible to read it as a defence against the accusation that he just got lucky himself. He elected to study tartaric and racemic acids because Mitscherlich had discussed them as an example of isomorphism. He could not have known that racemic acid is one of the very few chiral organic compounds that will spontaneously separate into its two enantiomers as it crystallizes – most other such compounds will forms crystals in which the two enantiomers are mixed together. Moreover, few other organic compounds will form crystals so large that they can be readily inspected with a nineteenth-century microscope. And even racemic acid requires the right conditions to separate in this way: it happens only at relatively low temperatures, and so it was lucky that Pasteur happened to be carrying out his experiments in an unheated Paris laboratory in February. Had it been July, the result might have been quite different.

Race for the racemate

In late 1848 Pasteur left Paris to take up a teaching position at the lycée in Dijon, but this seemed to him to be a remote backwater, and he eagerly accepted a subsequent offer from the University of

Strasbourg, where he began work in January 1849. He was sure that his discovery of what would later be called molecular chirality had yet more fruits to bear.

He asked Kestner at Thann for more racemic acid, which the industrialist duly provided; Kestner subsequently became Pasteur's trusted supplier and correspondent. Pasteur figured that Kestner's wine-making process first produced only right-handed tartaric acid, which subsequently became converted into racemic acid. But how could that happen? It meant that precisely half of the molecules had to acquire an opposite twist. 'It would be necessary', he said, 'to find an agent capable of producing this molecular modification on tartaric acid.' But how do you alter the twist of a molecule?

The problem was so profound that in 1851 the Société de Pharmacie in Paris offered a prize of 1500 francs to anyone who could explain the origin of racemic acid. The ambitious Pasteur was less interested in the money than in the fact that winning this prize would enhance his fame. So he set out to crack the puzzle.

During 1852 he travelled all over Europe, visiting laboratories and industrial plants to gather clues. Yet the quest seemed hopeless, and in January of 1853 Pasteur decided that the transformation of tartaric into racemic acid looked impossible. Only six months later, however, he had his answer.

He studied several other optically active organic compounds to look for a relation between molecular asymmetry and crystal form (hemihedry). The similarity between crystals of tartaric acid and those of an organic compound known as malic acids suggested to him that their molecular structures might be related. That is to say, they might both contain the same chiral 'core'. 'If there exists a common molecular grouping between the right-handed tartaric acid and the known malic acid of the mountain ash', Pasteur wrote,

> there is necessarily a common molecular grouping between the left-handed tartaric acid and a malic acid as yet unknown, and which would be to the malic acid known to chemists, what left-handed tartaric acid is to right-handed tartaric acid.

Maybe, then, nature made racemic acid not by reversing half of right-handed tartaric acid but by making both left- and right-handed forms from corresponding precursor molecules (such as malic acid) and then mixing them. But what about the precursors – where did *their* handedness come from? Pasteur came across an extraordinary claim by a chemist named Victor Dessaignes at Vendôme, who said that he had made malic acid from two other organic acids, maleic and fumaric acids. But those starting materials were optically *inactive*, whereas malic acid was optically active. 'Now to this day', Pasteur commented, 'no optically active substance has ever been prepared by laboratory procedures, starting with substances which are not optically active.' So could Dessaignes be mistaken?

Pasteur repeated his experiment, and found that it produced optically inactive malic acid. He concluded that this form of the compound was somehow 'untwisted': he called this mesomalic acid, introducing the 'meso' terminology. Now, as it happens, malic acid does, like tartaric acid, have an optically inactive meso form. But that's *not* what Pasteur made using Dessaignes' procedure – rather, he made a racemic mixture of left- and right-handed malic acid. Unaware of this, he set out to see whether he could make the tartaric acid analogue of what he deemed to be his mesomalic acid. During these studies, he found by chance that the optically active compound of tartaric acid, combined with a substance called cinchonia – an extract of the bark of the cinchona tree, which was the source of the antimalarial drug quinine – would, when heated, become transformed into racemic acid. By luck (and, of course, mental preparation) he had stumbled onto the solution that won him the Société de Pharmacie prize.

Chirality and life

These experiments hint at how Pasteur was starting to think about molecular chirality. At first, his claim that a chiral molecule could not be prepared from an achiral one rested simply on careful reasoning based on what was known about such chemical transformations (notice that his investigation of Dessaignes' claim did not in the end seem to him to contradict this notion). Gradually, however, this property of chirality became for Pasteur an article of faith. He was convinced that there was an impermeable barrier between chiral and achiral molecular forms – and, moreover, he thought that this same barrier separated the living from the inanimate world. Molecular chirality, he supposed, was uniquely a property of living systems.*

---

* The optical activity of minerals like quartz poses no real challenge to this supposition, since it was clear that this was not a *molecular* property – it results merely from the contingent arrangement of molecules in the crystal, and vanishes when the material is dissolved.

'Every chemical substance', Pasteur wrote,

> whether natural or artificial, falls into one of two major categories, according to a spatial characteristic of its form. The distinction is between those substances that have a plane of symmetry and those that do not. The former belong to the mineral, the latter to the living world.

This strong assertion touched on the protean notion of vitalism – the idea that there was a 'vital force' that quickened life, and which made living or organic matter fundamentally different from inorganic matter. By the mid-nineteenth century, the vitalistic hypothesis was looking distinctly embattled. But the debate was far from settled, despite later claims about the influence of Friedrich Wöhler's synthesis of urea in 1828 (see page 122). Pasteur was no vitalist – indeed, his own work on microbes helped to dispel the belief that a vital force caused the spontaneous generation of organisms in decaying matter – but he was keen to make a fundamental distinction between the organic and the inorganic, based on the idea of chirality.

Yet where did this handedness of life come from in the first place? Pasteur reasoned that it must be impressed on organic matter during its formation by the action of some all-pervasive chiral force. He illustrated how such a mechanical force might be manifested by analogy with a left- or right-handed screw being driven into a wooden block 'in which the fibres themselves have a right or left spiral arrangement'. In one case the screw would go in easily; in the other, only with difficulty. 'Is it not necessary and sufficient', he concluded, 'to admit that at the moment of the elaboration of the immediate principles in the vegetable organism [in other words, at the moment the molecules are formed], an asymmetric force is present?' Pasteur speculated that this force could be imposed by the environment, perhaps as an electric or magnetic field:

> Do such asymmetric agencies arise from the cosmic influences light, electricity, magnetism, heat? Do they perhaps stand in relation with the movement of the earth, with the electric currents by means of which physicists explain the earth's magnetic poles?

If so, then maybe these forces could be reversed, inverting the chirality of natural organic molecules. Consequently, in 1853 Pasteur embarked on a series of experiments that now look decidedly cranky: crystallizing substances in magnetic fields, growing plants from seeds irradiated with sunlight that was 'inverted' by reflection from mirrors. Biot implored him to abandon this strange quest, and indeed Pasteur himself admitted that 'One has to be a little mad to undertake what I am trying to do now.'

Yet chirality led him in more productive directions too. In 1854 he left Strasbourg to work at Lille, where there was an active brewing and distilling industry, and he was asked to help with some of the problems the industrialists encountered during the fermentation process. Fermentation was itself a disputed matter at that time. In 1837 Baron Charles Cagniard de la Tour proposed that the fermentation of sugar, by which means it is converted into alcohol, depends on the action of yeast: that is, of microscopic organisms. But others disagreed. The eminent German chemist Justus von Liebig suggested that yeast might be just a by-product of fermentation, and was not essential to what was basically a chemical process. Berzelius argued that the role of yeast was simply to 'awaken slumbering affinities' between the chemical compounds involved in fermentation: he gave this 'assisted' breakdown of substances a new name: catalysis. (Liebig's colleague Oerhardt complained that 'Calling the phenomenon catalytic does not explain it, it only replaces a name of ordinary language by a Greek word.')

Pasteur was drawn to study fermentation partly because one of the products, amyl alcohol, was chiral. This, he thought, could only be produced by the action of living organisms: fermentation could not be simply 'chemical'. He observed that the fermentation of sour milk into lactic acid – which happens spontaneously if milk is left to go off – produces tiny grey particles that, if extracted from the ferment, are then capable of converting sugar into lactic acid too. He decided that this 'lactic yeast' was a microorganism that multiplied as a result of fermentation, from which it derived sustenance. In other words, different kinds of microorganism gave rise to different modes of fermentation. This sowed the seeds for the discipline of microbiology, as well as paving the way to the discovery, at the end of the nineteenth century, of nature's molecular catalysts: enzymes, the key agents of biochemistry.

In late 1857, Pasteur studied the fermentation of tartaric acid when it was left to age. He found that, with racemic acid, only the right-handed isomer was fermented. This showed him that living organisms made precise distinctions between chiral molecules: 'We see here the character of molecular asymmetry peculiar to organic substances

intervene as a modifier of affinity [that is, of chemical reactivity].'
Pasteur discovered that our own physiology makes similar distinctions: one enantiomer of a chiral sugar may produce a sensation of
sweetness while the other barely registers. 'When active asymmetric
substances are involved in producing an impression on the nerves', he
said, 'their effort is translated by sweet taste in one case and almost no
taste in the other.' We now know that this exquisite discrimination in
molecular physiology is governed by the ability of enzymes to 'feel'
the difference between mirror-image molecular shapes. One consequence is that our bodies can only metabolize right-handed sugars,
which are the naturally occurring forms. Because our sweet taste buds
can nevertheless be stimulated by some left-handed sugars (such as
left-handed glucose), these are manufactured industrially as artificial
sweeteners, which give the taste without the calories of normal sugar.

The sensitivity of enzymes to chirality also accounts for nature's
enviable ability to generate just one of the two enantiomeric forms of a
chiral compound. Chemists today still find it a tremendous challenge
to design synthetic catalysts that can effect comparably selective
chemical transformations. Yet it is the very selectivity of the body's
biochemistry that creates a demand for such catalysts. While some chiral
compounds are simply physiologically inactive in one enantiomeric
form, others can elicit more troublesome responses. The most notorious
example is the drug thalidomide, which was administered to pregnant
women in the 1950s and 1960s to suppress morning sickness. While
the right-handed form acts as a sedative and an anti-inflammatory
agent, the left-handed form causes fetal growth abnormalities. Because
the marketed drug contained a mixture of the two enantiomers*, it led
to a wide incidence of birth defects in babies born in the early 1960s.

The thalidomide case made it clear that one must distinguish carefully
between the physiological effects of both enantiomers of chiral drugs,
and many drugs now have to be prepared in enantiomerically pure form.
This has led to the development of methods of separating enantiomers
(something that Pasteur also pioneered), although the ideal and most
economical solution is to make only one enantiomer in the first
place: hence the demand for enantiomerically selective catalysts (see
page 159). Several such manufacturing processes use natural enzymes,

---

* It is often implied that the adverse consequences of thalidomide would have been avoided if the
  drug had been enantiomerically pure – that is, if it had contained only the right-handed
  molecules. But in fact these get converted into left-handed molecules by an enzyme in the liver.

either extracted from microorganisms or within whole cells bred in fermentation vats, to conduct these delicate operations – for if Pasteur was wrong to think that only nature can generate chiral molecules, nevertheless it remains true that she is a lot better at it than we are.

Myths and Romances

Crucibles. Creations of Fire. The Last Sorcerers. Mendeleyev's Dream.
The titles of popular histories of chemistry reveal their intention to tell
a story whose roots are firmly bedded in legend, romance and reverie.
Even if we set aside the alchemists and their quixotic quest, chemistry
has a ledger of stories to rival the collection of Hans Christian Andersen.
Michael Faraday writing as a poor, self-tutored boy to beg Humphry
Davy for a humble position at the Royal Institution. Justus von Liebig
inviting Joseph Louis Gay-Lussac to celebrate an experimental triumph
by dancing around the laboratory. Friedrich August von Kekulé
dreaming up the structure of the benzene ring as he dozed before
his fire. And Dmitri Mendeleyev, as wild and exotic as any alchemist,
closing his eyes in exhaustion and seeing the periodic table take shape
in his head.

Many of these stories have, like that of Pasteur's crystals, now felt
the sharp scalpel of a historian's scepticism, and have duly collapsed
into a heap of hearsay and pure invention. But the real question is
why chemistry is so prone to these fancies – much more so, it seems,
than other science. That they continue to be repeated, more or less
uncritically, in popular books is not simply a result of indifferent
research: it is as though we *want* these stories to be true, as if they help
us make sense of chemical history.

The tradition of romanticization can be attributed clearly enough to
the chemists of the nineteenth century. Many of them were fully fledged
Romantics themselves, temperamentally allied to the strand of artistic
endeavour that connected Goethe, Wordsworth, Coleridge and J. M. W.
Turner. Davy himself was an archetypal Romantic – his Byronic good
looks and his elegant figures of speech had the society ladies of
London swooning at the Royal Institution. James Kendall, in his

canonically romantic *Great Discoveries by Young Chemists* (1938), portrayed him thus:

> No matinée idol of the last generation, no film star of the present day, ever created such a furore as this young 'Pirate of Penzance' when he first burst upon the delighted metropolis. "Those eyes were made for something more than poring into crucibles", said the fashionable ladies who swarmed to his lectures, and his desk was littered with anonymous sonnets from his fair admirers.

The curly-haired, dashing young man painted by Sir Thomas Lawrence could as well be Jane Austen's Mr Darcy; and Davy was no doubt fully conscious of the effect he made. He was not alone among chemists of his day in writing poetry, but he had unusually ambitious aspirations for it, and his friends Wordsworth and Coleridge thought well enough of Davy's verses to publish some of them in their *Annual Anthology* in 1799. Coleridge was famously said to have attended Davy's lectures to swell his own stock of metaphors.

Faraday was a simpler – some might say a less pretentious – man, but he was equally enthralled by the prevailing romanticism of the early nineteenth century. He had a keen eye for art, and befriended both John Constable and his arch-rival Turner. Sensitive to the Romantic vision of the sublime, he spent many happy days walking over the meadows, crags and glaciers of the Alps with his brother-in-law, the landscape artist George Barnard.

Before we deplore the embellishments, excesses and sheer fabrications in the accounts that the nineteenth-century chemists left of their research, however, we should remember that they generally saw nothing wrong in myth-making. It was widely accepted as a way of giving a story broader appeal, the equivalent of a modern journalist's poetic licence. Moreover, the fact that these scientists often drew on a stock selection of imagery implicitly acknowledges that their inventions were not necessarily meant to be taken at face value. The starchy Pasteur is, as I have said, one of the last people you would expect to rush out and prance down the corridor with a colleague, but he was merely employing the same trope as Liebig. Likewise, Kekulé and Mendeleyev both invoked an ancient association when they described their breakthrough moments as the products of dreams.

Mendeleyev's claim that the correct ordering of the elements came to him in a flash during an afternoon reverie seems evidently to have been a conscious retelling of history. Having written the symbols of the elements on a stack of cards, he had been shuffling and reordering them obsessively for three days and nights, he said. Late in the day on 17 February 1869, he dozed off through exhaustion. 'I saw in a dream a table where all the elements fell into place as required', he claimed. 'Awakening, I immediately wrote it down on a piece of paper.'

If only it had been so simple. Quite aside from the likely, if subliminal, influence of prior work on chemical 'families' by Johann Wolfgang Döbereiner, William Odling, Julius Lothar Meyer and John Newlands, Mendeleyev seems in fact to have arrived only gradually at his final ordering of the table and an appreciation of the significance of its periodic patterns. Historian Michael Gordin of Princeton University has questioned the whole idea of Mendeleyev's solitaire-like shuffling of cards.

Kekulé's dreams are even more contentious. He asserted that they supplied both of his major discoveries: the ring structure of benzene and the fourfold valency of carbon (the tendency of carbon atoms to form four chemical bonds each). Here is Kekulé describing the first of these dream visions, which came to him in 1854 as he drowsed on a bus carrying him to Clapham in London:

I fell into a reverie and lo, the atoms were gambolling before my eyes!... I saw how, frequently, two smaller atoms united to form a pair; how a larger one embraced two smaller ones... I saw how the larger ones formed a chain, dragging the smaller ones after them...

Compare this to Kekulé's dream of carbon rings at the fireside in Gent in the winter of 1861–2:

Again the atoms were gambolling before my eyes.... But look! What was that? One of the snakes had seized hold of its own tail, and the form whirled mockingly before my eyes.

Kekulé disclosed both of these dreams only in 1890, at a conference held in his honour in Berlin. Indeed, he spoke of the two events back to back in his address to the meeting, admitting their similarities, and one might have thought that he would expect this very juxtaposition to undermine the plausibility of his accounts (as it surely does). But it is possible that Kekulé was really just employing the images to make a rhetorical point, which he summed up directly afterwards: 'Let us learn to dream, gentlemen: then we shall perhaps find the truth.' All of this might be deemed harmless enough, even if it did rather confuse matters for subsequent historians who took Kekulé at face value. But it has also been suggested that the dream mechanism was a convenient way for Kekulé to overlook the work of others – for he remained forever reluctant to acknowledge that Archibald Couper independently conceived the idea of carbon's fourfold valency, or that Josef Loschmidt drew ring-like images of benzene before Kekulé proposed the notion.

The German chemist Hermann Kolbe did not approve of Kekulé's fanciful story, charging that Kekulé 'did not learn early to put his thoughts in order, to think logically and to restrain his fantasy'. Yet Kolbe was one of the prime culprits in launching another of chemistry's nineteenth-century myths: the demise of vitalism caused by Friedrich Wöhler's synthesis of urea from ammonium cyanate. As I have indicated (page 115), the idea that organic matter is imbued with some 'vital force' that distinguishes it absolutely from inorganic matter waned only gradually during the course of that century, due in part to Pasteur's demonstration that spontaneous generation of life does not occur. But it became standard for chemistry textbooks to claim that Wöhler's experiment sounded vitalism's death knell.

One of the earliest and most forceful depictions of the Wöhler myth occurs in Kolbe's chemistry textbook *Lehrbuch der organischen Chemie* (1854). Here he called the experiment 'epochal and momentous' and claimed that as a result of it, 'the natural dividing wall that separated organic from inorganic compounds came down'. This picture became increasingly absolute, as well as romanticized, and was eagerly promoted by Wöhler's biographers after his death in 1882. By the time it was picked up in Bernard Jaffe's immensely popular *Crucibles*, Wöhler was a Wagnerian figure on a glorious quest:

> He was standing upon the threshold of a new era in chemistry, witnessing 'the great tragedy of science, the slaying of a beautiful hypothesis by an ugly fact' .... The pregnant mind of young Wöhler almost reeled at the thought of the virgin fields rich in mighty harvests which now awaited the creatures of the crucible. He kept his head. He carefully analyzed his product to verify its identity. He must assure himself that this historic crystal was the same as that formed under the influence of the so-called vital force.

Crucibles provided the template for several subsequent histories of chemistry: as well as being highly romantic and over-dramatized, it exemplifies the Whiggish approach that professional historians have now long abandoned. In this view, all of scientific history is seen through the lens of the present, so that historical ideas are labelled 'good' or 'bad' depending on their congruence with contemporary scientific understanding. Needless to say, this tendency, which one can still find in some popular histories of science today, does few favours to the 'history' in such accounts; but all too often, telling history is not the objective. Rather, this form of science history has a triumphalist agenda that asks us to marvel at how far we have progressed beyond the ignorance and murk of former times. It says 'What a piece of work is man!'

But perhaps we should not be too harsh on the urge to romanticize and mythologize. Every discipline, according to science historian Peter Ramberg, needs its founding myths. 'The appeal of the Wöhler story, like many myths about origins', he says, 'lies perhaps in its ability to pinpoint the beginning of organic chemistry to a single datable event'. These myths are passed on to new practitioners as a kind of initiation ritual, like the 'invented traditions' that historian Eric Hobsbawm has identified in cultural life: holidays, monuments, sporting events. They foster a sense of community. 'Telling stories,' says Roald Hoffmann, 'is human and absolutely essential when we talk and write to and for each other. Story-telling provides psychological satisfaction, the drive to go on.' And we all need some of that.

CHAPTER 7

Life and How To Make It

Urey and Miller's Prebiotic Chemistry and the Beauty of Imagination

Chicago, 1952—Graduate student Stanley Miller persuades his distinguished supervisor, Nobel laureate Harold Urey, to let him carry out a bold and slightly crazy experiment: to simulate in the laboratory the kind of chemistry that might have gone on in the seas and lakes of the early Earth, before life colonized the planet. Urey suspects it is a fool's quest; but, within a week, Miller has generated the basic building blocks of life from a crude mixture of simple gases.

Everyone knows that the simplest questions are the most difficult to answer, but sometimes it is hard even to know where to start. Confronted with the puzzle of how geology gave birth to evolution, Charles Darwin appeared to throw up his hands in despair in 1863. 'It is mere rubbish', he said in a letter to his friend Joseph Hooker, 'thinking at present of the origin of life. One might as well think of the origin of matter.'

Scientists are now thinking very hard about the origin of matter, and although in some ways the answers remain as remote as ever, nevertheless it is fair to say that the question has been considerably refined. But what about the origin of life?

The beauty of the experiment performed by Stanley Miller, under Urey's supervision, in 1952 is that it displayed the kind of imagination without which one cannot start to open up a truly deep question. Miller dared to portray this question in impressionistic terms, despite the fact that science is generally supposed to be precise, hard-edged, tightly focused. He didn't worry that the question was not even understood well enough to frame a definite hypothesis that one might then take to the experimental testing ground. He didn't know quite *what* to expect from his experiment. And yet still the experiment was meaningful – sufficiently so, in fact, for Miller's results to prove shocking.

Perhaps he was simply too young to appreciate the gamble he was taking. Urey was too experienced to believe, at first, that it was worth the risk. They must both, though, have instinctively sensed that well-educated guesswork might be enough, in this instance, to derive real science from what seemed like a ridiculously crude and speculative enterprise. Before the Miller–Urey experiment, studying the chemical origin of life was a kind of parlour game for iconoclasts and grand old men; afterwards, it was an experimental science.

Quite aside from the insights the experiment gives us about the way life began (and those implications are still disputed today), the work that Stanley Miller did in Chicago can offer inspiration to scientists facing the impossible. For the origin of life does indeed seem a wholly implausible event, as biologist George Wald confessed in 1954: 'One has only to contemplate the magnitude of this task to concede that the spontaneous generation of a living organism is impossible.' Yet here we are, he added – the impossible made flesh. Nature evidently succeeded, and so there must be an answer to the conundrum. The daunting task for science is to find it.

Miller and Urey did not, of course, come anywhere close to generating an organism spontaneously in a test tube, although after reading some of the contemporary reports in the media you might be forgiven for thinking that they had. But suddenly such a feat seemed a whole lot *less* impossible: it was, you might say, the 'impossible' of an exceedingly hard problem, not the 'impossible' of a transgression against nature. Many scientists now suspect that it will be only a matter of a decade or so before an organism that is in all meaningful respects 'living' will indeed have been fabricated 'from scratch', from its basic molecular building blocks, put together in the laboratory. That won't in itself answer the question of how life began, but it will surely owe some kind of debt to the discovery by Miller and Urey that life's fundamental chemistry is in some ways more accessible than we might think.

Searching for origins

Charles Darwin evidently changed his mind after writing to Hooker, for by 1871 he could be found thinking himself about the origin of life. 'It is often said', he mused, that all the conditions for the first production of a living organism are now present, which could ever have been present. But if (and oh! what a big if!) we could conceive in some warm little pond, with all sorts of ammonia and phosphoric salts, light, heat, electricity, etc., present, that a proteine compound was chemically formed* ready to undergo still more complex changes, at the present day such matter would be instantly devoured or absorbed, which would not have been the case before living creatures were formed.

By then, chemists and biochemists were already starting to take the question more seriously. Even if Friedrich Wöhler's synthesis of urea in 1828 (see page 122) was not exactly the blow for vitalism that later historians made of it, nonetheless it makes the point that organic materials can come from inorganic ingredients. In addition, careful work on the basic chemical processes of life, such as fermentation and respiration, in the mid-nineteenth century prompted researchers to wonder how these biochemical reactions might have first come about. In 1875, physiologist Eduard Pflüger's researches on 'physiological combustion in living organisms' led him to suspect that the primeval generative substance of life on Earth was the cyanide free radical, or its paired-up molecular form cyanogen, $(CN)_2$.

Pflüger identified various inorganic reactions that could produce cyanogen, which in turn could give rise to the key ingredient of living cells: protein. 'One sees', he said,

how quite extraordinar[il]y and remarkably all the facts of chemistry indicate to us fire as the force which has produced the constituents of protein through synthesis. Life descends thus from the fire and is in its fundamental conditions generated at a time when the Earth was still a glowing ball of fire. If one considers now the immeasurably long time in which the cooling of the Earth's surface took place – infinitely slowly, so cyanogens and the compounds which contain hydrocyans and hydrocarbons, had sufficiently long time and opportunity to follow their great inclination towards transformation and formation of polymers in the most extensive and different ways, and with participation of oxygen and later of water and the salts to go over into that self-decomposing protein which is living matter.

---

* Darwin's chemistry seems a little confused here: there's no need for 'phosphoric salts' to make 'proteine'. Yet curiously, something of that kind does appear essential for putting together nucleic acids such as DNA and RNA, of which Darwin would have had no inkling.

This remarkable passage summarizes all the important issues in the problem of life's origin. Pflüger wonders what the early Earth looked like, and what kinds of 'prebiotic' chemistry were possible on its surface. He proposes, as did Wald 80 years later, to solve the problem of impossibility by appeal to the vastness of geological time – the same solution, of course, that Darwin invoked to explain how humankind came out of the primordial slime. He acknowledges that, for life to begin, small, simple building blocks must form polymers, and that these must ultimately organize themselves into living systems.

But what were conditions truly like on the early Earth? Could life have arisen when it was indeed a 'glowing ball'? Or does there have to be liquid water available? Also, was there, as Pflüger assumes, oxygen in the atmosphere in those distant days?

This latter question was considered in the 1920s by a young Russian biochemist named Alexander Oparin. In 1924 he wrote a pamphlet in which he drew on the available astronomical observations of the chemistry of stars, comets and meteorites to deduce what the atmosphere of the early Earth was really like. 'The atmosphere of that time', he concluded, 'differed in many respects from that of today. Water vapour was especially abundant in it.' Oparin believed that reactions between water and the compounds formed from metals and carbon on the planet's surface would have produced an atmosphere rich in 'compounds of carbon with hydrogen and oxygen' – particularly hydrocarbons. 'These were the first "organic" compounds on the Earth', he concluded. They might have reacted with oxygen to supply a rich mélange of organic substances, such as alcohols, ketones and organic acids. Moreover, Oparin figured that ammonia would be created by the reaction of nitrogen with superheated steam, and he allowed that Pflüger's cyanogen molecules might have been present too. His main point was that the young planet, far from being a barren rocky wilderness, could have been rich in organic molecules, providing not only the building blocks of primitive life-forms but also their sources of nutrition.

Oparin's discussion introduced another idea that later became central to the Chicago experiment. In his day, scientists had figured out that the early atmosphere was indeed probably very different from today's – most scientists thought it would have been composed primarily of carbon dioxide (CO_2) and nitrogen (N_2). In contrast, Oparin talked about the prevalence of gases such as hydrocarbons – methane (CH_4), for example – and ammonia (NH_3). Here the carbon and nitrogen atoms are bound to hydrogen. That is a very different kind of mixture from a CO_2/N_2 atmosphere. A mixture containing a prevalence of hydrogen, whether in compounds with other elements or in elemental form (H_2 gas), is said to be *reducing*, while one that contains oxygen-rich molecules is called *oxidizing*. In crude terms, an oxidizing atmosphere has a tendency to burn up organic compounds. Our present-day atmosphere, one fifth pure oxygen, is highly oxidizing, which is why wood and plastics and organic solvents are apt to go up in flames, and why alcohol gains oxygen and turns to vinegar when exposed to air. A reducing atmosphere is much gentler on organic compounds. Oparin raised the notion that the early atmosphere might have been reducing rather than oxidizing.

He made this point more forcefully in the 1930s, after astronomers discovered that the atmospheres of the gas-giant planets Jupiter and Saturn were composed largely of hydrogen, methane and ammonia. If these gaseous mantles were acquired when the planets were formed, so that they represent a sample of the gaseous stuff from which the solar system condensed, wouldn't the smaller, rocky inner planets have started off with atmospheres like this too?

In the meantime, the British biologist J. B. S. Haldane, quite unaware of Oparin's work, began to formulate his own ideas about the primeval environment that nurtured the first life on Earth. He decided that the early atmosphere lacked oxygen, and so he proposed that the first organisms were anaerobic. He envisaged an atmosphere of carbon dioxide, water vapour and ammonia, and he argued that, without oxygen, there would be no ozone layer to filter out the harsh ultraviolet rays of the sun. This high-energy radiation would have cooked up all manner of organic compounds in the water on the planet's surface, turning the oceans into a 'hot dilute soup'. Yet in 1929 Haldane admitted that his ideas were speculative, and that

> they will remain so until living creatures have been synthesized in the biochemical laboratory. We are a long way from that goal.... I do not think I shall behold the synthesis of anything so nearly alive as a bacteriophage [a bacterial virus] or a virus, and I do not suppose that a self-contained organism will be made for centuries. Until that is done the origin of life will remain a subject for speculation. But such speculation is not idle, because it is susceptible of experimental proof or disproof.

Scientists were in fact already hunting for such 'experimental proof or disproof'. By conducting experiments on mixtures of simple gases and water subjected to a source of energy to induce chemical reactions, they were seeking to create a 'cartoon' version of the prebiotic Earth. Researchers at the University of Liverpool in England showed that Haldane's mixture of water, ammonia and carbon dioxide could be converted into organic compounds by irradiation with UV light.

Experiments of this sort had commenced in the mid-nineteenth century, when chemists, following Wöhler's lead, found ways of converting simple compounds into the molecules of life. In 1850 Adolph Strecker showed that ammonia, hydrogen cyanide and the organic compound acetaldehyde (produced when ethanol is oxidized by air, and responsible for the bitter taste of wine that has 'gone off') can be combined to make the amino acid alanine. (Proteins are polymers made by linking together 20 varieties of amino acid into long chains) Eleven years later the Russian Aleksandr Butlerov made sugar molecules by treating formaldehyde with caustic soda. As well as providing metabolic fuel, sugars are the building blocks of carbohydrate polymers such as starch and cellulose, and are components of nucleic acids. But these scientists had no intention of exploring the origin of life: they considered that they were simply carrying out synthetic organic chemistry.

In 1906 the biochemist Walther Löb used electricity to energize mixtures of carbon dioxide, carbon monoxide, ammonia and water. He wasn't seeking to study prebiotic chemistry either – instead he was trying to mimic the physiological processes by which plants convert 'inorganic' sources of carbon and nitrogen into useful biomolecules. Nonetheless Löb found that he could make some surprisingly sophisticated organic compounds this way, including aldehyes such as formaldehyde and the amino acid glycine.

A Mexican scientist named Alfonso Herrera set out explicitly to explore the chemical origin of life in the 1940s, and he found that two amino acids can be made from formaldehyde and cyanide salts. Other researchers succeeded in synthesizing fatty acids and sugars from simple starting materials, for example using electrical discharges to drive the reactions. Typically, however, these experiments on prebiotic mixtures remained wedded to the idea that the early atmosphere was oxidizing, and so they used ingredients such as carbon dioxide and nitrogen. The results were uniformly disappointing: the researchers rarely succeeded in making anything more exciting than formaldehyde, and even then in tiny amounts. The studies conducted in the early 1950s at the University of California at Berkeley by Melvin

Calvin – who later won a Nobel for elucidating the way plants take up carbon dioxide from the air – were somewhat typical. Calvin bombarded mixtures of carbon dioxide, hydrogen, and a solution of iron salts with high-energy helium ions produced by the Berkeley particle accelerator. He found nothing more suggestive than a bit of formaldehyde and formic acid. In Chicago, however, there were other things afoot.

Cooking the primal soup

Harold Urey was world-famous when Stanley Miller came to Chicago. In 1931 Urey discovered heavy hydrogen (deuterium), and this work earned Urey the Nobel prize in chemistry in 1934. During the war he applied his chemical skills to the separation of uranium isotopes for the Manhattan Project (see page 72). He came to the Institute for Nuclear Studies in Chicago in 1945, where Enrico Fermi, the architect of the first nuclear reactor, held court.

Urey's interest in isotopes led him into the field of radiochemical dating of geological materials, which in turn awakened an interest in the origin of the Earth and the solar system. Without knowing about Oparin's ideas, Urey came to the conclusion in 1951 that the early Earth must have had a reducing atmosphere. In October he gave a talk in which he suggested that someone should try to mimic prebiotic chemistry under such conditions. Miller, Urey's graduate student, was in the audience.

'So I went to him and said "I'd like to do those experiments"', Miller later explained. 'The first thing he tried to do was talk me out of it.' Urey thought that the chances of finding anything interesting this way were slim, and that it would take a long time, and that this would not be a good way for Miller to gain a doctorate. But Miller persisted, and eventually Urey relented. 'We agreed to give it six months or a year', Miller says.

It was autumn of 1952 before Miller was ready to begin the work. The simple equipment that he and Urey devised is now iconic (Figures 19 and 20). Two tungsten electrodes converge in the upper part of a glass globe that holds five litres of gases, and a spark that arcs at regular intervals between the electrodes simulates lightning forking through the prebiotic atmosphere. The gases, including any organic molecules produced by the electrical discharge, pass down through the bottom of the globe and into a water-cooled condenser. Here, water

Figure 19 *Stanley Miller with the apparatus in which he cooked up amino acids from a mixture of simple gases*
(© CORBIS)

vapour in the mixture condenses to a liquid, which runs into a smaller flask that is heated to drive steam back into the large globe. As the mixture of gases circulates, products of the spark-induced reactions accumulate in the water. It is the very economy of the design that makes this experiment a plausible proxy for the early Earth: since we don't know what that environment was really like, the chances are that a more complex apparatus would fail to make the experiment any more realistic, but could merely render it harder to interpret and reproduce (such duplication happened rapidly after Miller and Urey published their results). 'The fact that the experiment is so simple that a high-school student can almost reproduce it', says Miller, 'is not a negative at all. The fact that it works and is so simple is what is so great about it.'

All the same, one can't help suspecting that this arrangement, with its gleaming glass bulbs and its ominous electrodes crackling with energy, unconsciously connected the experiment with those studies in the electrification of life that began in the 1780s with Luigi Galvani's

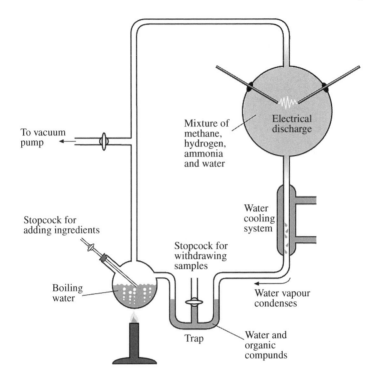

Figure 20   *The Urey–Miller experiment. The mixture of gases is energized by a spark in the globe on the top right, and reaction products are collected in the water that condenses and runs into a flask at the bottom*

jerking frog legs and culminated in Mary Shelley's Promethean fable in which Victor Frankenstein aims to 'infuse a spark of being into the lifeless thing that lay at my feet.' There was surely a Faustian frisson pervading the news that Miller and Urey had used electricity to conjure the raw molecules of life out of tenuous vapour.

That, however, was still to come. Heeding Urey's conviction that the prebiotic atmosphere was a reducing blend, Miller elected to experiment with a mix of methane, ammonia, hydrogen and water. In fact he devised three different experiments: while the electric-discharge system was the most successful, Miller also used an arrangement in which the steam rose under pressure directly into the discharge bulb, forming a hot mist that mimicked the environment of a water-rich volcanic eruption, and an apparatus in which the electric spark was replaced by a so-called silent electrical discharge.

Not even the young, optimistic Miller could have anticipated how quickly his flasks would cook up something interesting. Overnight, the water turned red: a sure signature of complex chemicals in the broth. After two days, he tapped off some of the brownish liquid and analysed it using the technique of paper chromatography. This involved placing a drop of the liquid on a piece of blotting paper that was then dipped in an organic solvent. As the solvent is sucked up between the fibres of the paper by capillary action, it carries the compounds in the sample with it – but they rise to different heights, depending on their chemical composition. Thus, the various substances in the mixture become separated. The separation can be improved by turning the paper by 90° and dipping it in a second solvent. If the components of the mixture are coloured, each of them becomes visible as a distinct blob deposited at a specific position on the paper sheet. Miller made his products more clearly visible by spraying the paper with a chemical that turned dark when it reacted with the different compounds present. Straight away he was able to identify the amino acid glycine, the simplest of the basic building blocks of proteins. After a week of running the experiment – by which time the sparking bulb had become coated with an oily yellow deposit – several other products were evident, including various amino acids.

In the end, Urey and Miller identified no fewer than 13 of the 20 amino acids found in natural proteins. And there was plenty of them. 'We would have been happy if we got traces of amino acids', Miller says, 'but we got around four percent'. In total, 10–15% of the carbon in the methane gas was converted into organic compounds within Miller's apparatus.

Urey and Miller could not contain their excitement, and they began to discuss the results at once in scientific talks. It is said that when Miller spoke at a seminar in Chicago, Enrico Fermi was in the audience and asked a little sceptically (Fermi was famously sceptical) whether this kind of process could really have happened on the early Earth. Urey came to his student's rescue, replying that if it didn't, then God missed a good bet.

The mass media had no such doubts. When Urey spoke about the experiments in November 1952, the news was picked up by both *Time* and *Newsweek*. After that, it was clearly imperative for the two researchers to publish a scientific paper as soon as possible so that other scientists could assess the findings. Urey knew that if his name appeared on the paper, everyone would credit the work to him and

Miller would be forgotten; so he insisted that Miller should be the sole author. But that did not prevent him from leaning on his reputation. When Miller submitted his paper to the journal *Science* in February 1953, Urey accompanied it with a personal call to the editorial offices to make sure they appreciated the importance of the results.

All the same, the manuscript had to undergo the peer review process. *Science* had had the paper for less than two weeks before Urey began to lose patience, and he wrote to Howard Meyerhoff, the chairman of the editorial board of the American Association for the Advancement of Science (which publishes *Science*), threatening to withdraw the paper and to submit it instead to the *Journal of the American Chemical Society* if they didn't get on and publish it soon. (Nobel laureates do this kind of thing all the time.) Miller is convinced that without Urey's backing the paper wouldn't have received serious consideration from the outset – 'it would still be on the bottom of the pile', he says.

But one of *Science*'s referees wasn't sure that the results were credible. After all, no one had made such a rich and abundant medley of prebiotic chemicals before. (Miller says that this reviewer later identified himself and apologized for the delay; but one can understand his caution.) Miller was still awaiting the referees' verdict a week later when he was dealt a crushing blow. On 8 March the *New York Times* ran a story describing work conducted by Wollman M. MacNevin and his colleagues at Ohio State University. It described very similar electric-discharge experiments using methane, which yielded 'resinous solids too complex for analysis.' Despite the lack of information about the chemical products, MacNevin's experiment was advertised as showing how complicated organic molecules could have arisen in Earth's early history. 'Their work', Miller told Urey disconsolately, 'is, in essence, my thesis.'

The prospect of being scooped galvanized Urey – on Miller's behalf he told *Science* that he was sending the paper instead to the *Journal of the American Chemical Society*. But Meyerhoff knew a good story when he saw one. He quickly wrote to Miller saying that he wanted to publish the paper as the lead article, and asking Miller to confirm that Urey's wishes reflected his. Miller decided that this kind of exposure in *Science* was the best way to publicize his achievement, and the paper appeared in that journal two months later. It was a revelation. 'If their apparatus had been as big as the ocean, and if it had worked for a million years, instead of one week', said *Time*, 'it might have created something like the first living molecule.'

'Living molecules' were very much in the air at the time, for only a few weeks previously James Watson and Francis Crick had described in *Nature* the structure of DNA, the 'molecule of life'. It must have seemed as though, all at once, the deepest secrets of creation were being revealed to the world.

Life itself

Chemical simulations of the prebiotic environment like those of Miller and Urey were later shown to be capable of new wonders. In 1959 Wilhelm Groth and H. von Weyssenhoff at the University of Bonn showed that ultraviolet light, rather than electric discharges, could induce the formation of amino acids in reducing mixtures, obviating the charge levelled at Miller and Urey that the early Earth would have had to have experienced continual lightning storms to make sufficient quantities of these compounds. In addition, two years later Juan Oró at the University of Houston in Texas showed that a mixture of hydrogen cyanide and ammonia in water will produce not only amino acids but adenine, one of the four key components of DNA. Soon after that, Sri Lankan chemist Cyril Ponnamperuma showed that a beam of high-energy electrons can generate adenine from a Miller–Urey mixture of gases.

Since then, chemists have devised ingenious ways of making all the key building blocks of life's molecules – amino acids and nucleotides, the basic structural units of DNA and RNA – from simple precursor molecules under conditions that are more or less plausibly 'prebiotic'. They have also succeeded in getting some of these molecules to link up into polymer chains of the sort found in proteins and nucleic acids. It could seem as though they are on the threshold of putting together life from the raw constituents of the solar system.

But that is not really the case at all. For one thing, some of these building blocks (particularly sugars) are not terribly stable in water: they tend to get split apart ('hydrolysed') so rapidly that it is hard to see how there would have been time enough to make complex molecules from them. Some of the other molecular building blocks require supposedly 'prebiotic' conditions that seem rather contrived. Furthermore, it's not at all obvious how they could have got concentrated enough, in the early oceans, to react together and lead to an increase in molecular complexity. Perhaps Darwin's 'warm little pond' is indeed what is needed there: a lagoon or lake baked by the sun so that evaporation creates an ever richer soup.

More problematically from the point of view of the Miller–Urey experiment, there is now evidence that the Earth's early atmosphere may not have been reducing after all, but mildly oxidizing. This remains contentious, and Miller himself is sceptical. If the first organic molecules were indeed formed by reactions in the Earth's atmosphere and oceans, he says, then his experiments show that the environment would *have* to have been reducing: 'Either you have a reducing atmosphere or you are not going to have the organic compounds required for life.... I have not found an alternative to disprove the need for a primitive reducing atmosphere.'

But, perhaps, some researchers say, the molecules of life weren't made this way at all. Some meteorites are rich in organic compounds, apparently formed through chemical reactions in interplanetary space or on the worlds from which the meteorites came. Louis Pasteur studied one such 'carbonaceous' meteorite, which fell near Orgueil in southern France, in 1864, to see if it contained bacteria (it didn't). Another one, which fragmented over the small town of Murchison in Australia in 1969, has proved a particularly rich source of complex organic molecules, including many amino acids. The Earth has apparently been showered with these substances since its earliest days, and some researchers believe that the principal source of life's building blocks could have been this extraterrestrial delivery, rather than formation *in situ*. Again Miller remains unconvinced that comets and meteorites would have carried enough organic material to Earth to sow the seeds of life; but it is widely viewed as a possible alternative.*

Another school of thought proposes that life arose not in the shallows of the early oceans but in their depths, where hot, mineral-rich water gushes out from so-called hydrothermal vents on the sea floor. These vent fluids are infused with volcanic methane, as well as ammonia, carbon dioxide and other potential ingredients of prebiotic chemistry. The volcanic heat itself supplies energy to drive chemical reactions. Hydrothermal vents support thriving ecosystems today at depths far too great to allow organisms to use sunlight for photosynthesis. Miller insists that these miniature subsea volcanoes have a destructive capacity that outweighs any potential creativity: he says

---

* There has been a much-disputed claim that some of the amino acids in the Murchison meteorite show a predominance of left-handed over right-handed enantiomers. This suggests the possibility that the exclusive use of left-handed amino acids in proteins (see Chapter 6) might have been 'seeded' by a bias in the extraterrestrial building blocks. But the possibility of sample contamination by terrestrial organic molecules is very hard to eliminate in such studies.

that organic molecules simply get burnt up in the scalding water. Again the matter is currently unresolved; but one can't help but be struck by Miller's determination to defend the thesis that he concocted in 1953 against all comers. It is understandable, perhaps, that someone whose 'crazy' experiment produced such remarkable results should not give up their fruits without a battle.

Quite aside from the provenance of the first complex organic molecules, the challenge of getting from a prebiotic soup to the first living cell remains so immense that researchers in this area still labour under the shadow of Darwin's early despair. Life is not a bag of chemicals – it is a process orchestrated with awesome precision and sophistication, in which all the parts seem to fit together in a truly wondrous design. In that sense, making the prebiotic building blocks is the easy part.

All the same, scientists have made a promising start on the awesome problem of how these molecules could have become assembled into cells. One of the most fertile ideas, devised in the late 1960s by Carl Woese, Leslie Orgel and Francis Crick, is the hypothesis that biologist Walter Gilbert christened the 'RNA World' in 1986. This is a way to solve the conundrum posed by the mutual dependence of DNA and proteins: without proteins, DNA cannot replicate or function as the repository of genetic information, but without DNA there is no blueprint for manufacturing proteins, each with its precisely defined structure. RNA, which acts as the go-between in turning the information encoded in DNA into protein catalysts (enzymes), might, in the RNA World, have performed both roles: gene bank and catalyst. The discovery in the 1980s that RNA molecules can indeed act as catalysts in cells today made it seem considerably more likely that the very earliest 'proto-life' went through a stage that looked something like the RNA World.

Although this all remains sketchy, advances in biotechnology and research in molecular biology and genetics have brought scientists to the brink of being able to build simple organisms from scratch. While we can never really know what went on nearly four billion years ago on planet Earth, we might now at least be able to explore some of the possible scenarios in the laboratory and see how they play out. In addition, there can be no doubt that this exciting and indeed rather frightening possibility of synthesizing life owes its greatest debt to the leap of imagination taken in 1952 by Stanley Miller and Harold Urey in Chicago. 'It was the Miller experiment', say 'exobiologists' Jeffrey Bada and Antonio Lazcano, 'that almost overnight transformed the study of the origin of life into a respectable field of enquiry.' Carl Sagan, himself one of the pioneers in the study of life in the universe, which now goes by the name of astrobiology, saw even broader implications in that experiment: it gave people like him the faith to believe that it is worth scanning the skies for signs that we are not alone. It was, Sagan said, 'the single most significant step in convincing many scientists that life is likely to be abundant in the cosmos'.

CHAPTER 8

Not so Noble

Bartlett's Xenon Chemistry and the Beauty of Simplemindedness

Vancouver, Canada, 1962—A young chemist at the University of British Columbia named Neil Bartlett demolishes a long-held conviction by demonstrating that xenon, an 'inert' gas, can after all combine with other elements to form a chemical compound. Bartlett's bright yellow-orange substance, with the chemical formula $XePtF_6$, becomes the first compound to be associated with this group of elements, and opens a new chapter in inorganic chemistry.

The gaseous elements in the right-hand tower of the periodic table, variously known as 'inert', 'noble' and 'rare', stand aloof from the promiscuous couplings of the other elements. Conventional wisdom long asserted that they felt no desire to make unions (it's all but impossible to resist getting a little anthropomorphic here). But maybe, H. G. Wells suggested in 1898, there are other beings who know better. In *The War of the Worlds*, invaders from Mars employ a more sophisticated chemistry than ours, enriched by elements yet unknown on Earth, in their plans for interplanetary conquest:

> The Martians are able to discharge enormous clouds of a black and poisonous vapour by means of rockets. They have smothered our batteries, destroyed Richmond, Kingston, and Wimbledon, and are advancing slowly towards London, destroying everything on the way. It is impossible to stop them. There is no safety from the Black Smoke but in instant flight.

What is this toxic Black Smoke? In the book's epilogue, after the Martians have been defeated by the common cold, the narrator confesses that scientists are still puzzled by it:

> Spectrum analysis of the black powder points unmistakably to the presence of an unknown element with a brilliant group of three lines in the green, and it is possible that it combines with argon to form a compound which acts at once with deadly effect upon some constituent in the blood. But such unproven speculations will scarcely be of interest to the general reader, to whom this story is addressed.

Indeed, the general reader could be forgiven an ignorance of argon altogether, since its discovery had only been announced four years earlier, and Wells' reference to it shows just how avidly he followed the latest scientific news. As we saw in Chapter 2, Lord Rayleigh and William Ramsay captured this elusive element after following up Henry Cavendish's puzzling comment about the composition of air. They knew that argon was inert – that, after all, was why they named it the 'lazy' element – but they could not know at that point just *how* lazy it would prove to be. In 1896 Ramsay wrote,

> It cannot, of course, be stated with absolute certainty that no elements can combine with argon; but it appears at least improbable that any compounds will be formed.

Others were not so sure about that. Ramsay sent a sample of argon to his friend, the eminent French chemist Henri Moissan, who suspected that it might combine with fluorine, the highly reactive element he isolated in 1886. Moissan sent sparks through a gaseous mixture of argon and fluorine, but found nothing, confirming Ramsay's impression that this was a profoundly inactive substance. Marcellin Berthelot looked for reactions between argon and benzene gas, and even believed he had found them, though Ramsay was sceptical. It soon became clear that argon was a deeply solitary element, inert in the face of anything chemists could throw at it.

And not just argon. All the other related gases that Ramsay and his colleague Morris Travers sifted from liquid air – neon, krypton and xenon – proved resolutely inert. By 1924 the Austrian chemist Friedrich Paneth pronounced what was essentially the consensus view of chemists worldwide: 'the unreactivity of the noble gas elements belongs to the surest of all experimental results'.

Forty years later the general opinion had hardly changed, which is why it took great boldness to do what Neil Bartlett did in 1962. It was, however, a boldness born of straightforward, textbook reasoning. The beauty of the experiment stems from that simplicity, that recognition and embracing of opportunity – that ingenuous leap into the unknown.

Moments of inertia

The inertness of the noble gases made such a neat story that it seemed beyond dispute. The reason for the inactivity of these elements fell out of the quantum-mechanical explanation for the periodic table and for chemical bonding that was developed in the 1920s. When quantum theory was applied to the way electrons are distributed around atomic nuclei, it showed that they form distinct 'shells', each with a characteristic number of electrons. The first shell has space for just two electrons, the next accommodates eight, subdivided into sub-shells of two and six, and the third shell contains 18 electrons, in sub-shells of two, six and ten. A shell structure had already been posited in the 1910s and 1920s by the American chemists Gilbert Lewis and Irving Langmuir, who perceived that it gives an empirical explanation for the way each element has a particular chemical-bond-forming capacity or 'valency'. The physicists Niels Bohr and Arnold Sommerfeld rationalized these ideas by showing that the electron-shell structure of atoms emerges directly out of quantum theory.

In the simplistic yet serviceable picture presented in school chemistry, chemical bonding arises from the tendency of atoms to 'seek' a fully filled outer shell (or sub-shell) of electrons by sharing them with other atoms. Each bond is forged from a pair of electrons, one from each of the two atomic partners in the union (bonds like this are called *covalent bonds*). Thus the carbon atom, with four electrons in a shell whose total capacity is eight, forms four chemical bonds to other atoms, thereby acquiring four other (shared) electrons to complete its electron shell. Ultimately, the driving force behind this formation of bonds is the lowering of energy: collectively, the atoms become more stable when they are bonded together. Four hydrogen atoms and one carbon atom have a lower total energy if they unite to form a methane molecule (CH_4) than if they remain as separate atoms. There are exceptions and complications to this idealized notion of chemical bonding as the formation of electron pairs and the filling of electron shells; and formulating a proper theory of it requires an injection of quantum mechanics, which is what the American chemist Linus Pauling provided in the 1930s. But, on the whole, this simple model describes the state of affairs reasonably enough.

Not all chemical compounds come about from electron sharing, however. Some elements gain a complete outer shell of electrons merely by casting off one or more electrons. This leaves the atoms with an overall positive electrical charge: they are positive *ions*. Metals tend to do this, especially those on the far left of the periodic table. The alkali metals, such as sodium and potassium, gain a full outer shell by losing a single electron, and this is what happens to sodium when it forms a salt such as sodium chloride. By the same token, some atoms gain a filled shell by acquiring extra electrons: a chlorine atom has seven electrons in a shell with a capacity of eight, and so it 'fills the gap' by picking up an electron from sodium. This turns the chlorine atom into a negatively charged ion. The oppositely charged sodium and chloride ions then attract one another, and stick together to form *ionic bonds*, which bind the ions into crystals of sodium chloride.

So the formation of chemical bonds, whether they are ionic or covalent, involves some change in the number of electrons that can in some sense be assigned to each atom, dictated by the tendency of atoms to acquire filled shells (or sub-shells) of electrons. As one progresses along each row of the periodic table of elements, each successive element is constituted of atoms with one more proton in the nucleus and one more electron 'in orbit' around it. So the shells are progressively filled with electrons from left to right. By the time we reach the inert gases, sitting at the end of each row, there are no more 'free spaces' to be filled by electron-sharing: every sub-shell occupied by electrons is already completely full. So there is no incentive for the inert-gas atoms to form covalent bonds.

What about knocking electrons out of the shells, or adding them to empty shells, to form ionic compounds? The inert gases don't lend themselves to that either. The electrons in each successive shell have higher energies, which is to say, the atom grasps them somewhat less tightly. Neon, say, has a full second shell, and to turn it into a negative ion by adding an electron would mean putting this extra electron in the third shell. The neon atom just can't hold onto such an electron tightly enough. (Much the same applies to sodium, whose nucleus is scarcely any bigger than neon's, but which *does* have a single third-shell electron. So it loses this electron at the drop of a hat, forming an ion – that is why sodium metal is so reactive.) How about turning an inert-gas atom into a *positive* ion by removing an electron from the outermost filled sub-shell? There is nothing in principle to prevent this; but it costs a lot of energy. Roughly speaking, the electrons in an atom's outer shell get held increasingly tightly from the left to the right of a row in the periodic table, because the positive charge on the nucleus (which is what binds the electrons in the atom) increases along the row. So by the time we reach the inert gases, the electrons are pretty strongly bound and hard to remove.

That is all a way of saying that the configuration of electrons in inert gas atoms is especially stable just as it is, so that these elements look set to resist any attempt to perturb the arrangement of electrons by forming compounds. As a consequence, the atoms have very little propensity to stick together: at everyday temperatures and pressures, the elements are all gases in which the atoms drift about alone.

This did not prevent some scientists from speculating in the early days of quantum chemical theory that the inert gases *might* form stable crystalline compounds. In 1916 Walther Kossel suggested that a sufficiently 'electron-hungry' element, like fluorine or oxygen, might just be capable of stripping an electron from the heavier inert gases, allowing them to form ionic compounds. (The bigger and heavier the atoms, the less tightly their outermost shell of electrons is bound.) Andreas von Antropoff in Germany said much the same thing in 1924 – he even thought, wrongly, that he'd made a compound of krypton and bromine. Eight years later Linus Pauling himself speculated that fluorine could conquer the solitary nature of an element like xenon. He begged a sample of xenon (which was still hard to come by in those days) and took it to his colleague, chemist Don Yost at the California Institute of Technology. Yost and his student Albert Kaye laboured valiantly to make a compound of xenon and fluorine using Moissan's approach of passing electrical discharges through a mixture of the gases. But nothing came of it – all they obtained was the compound silicon tetrafluoride, as the virulent fluorine attacked the quartz walls of the container.

A simple idea

At first, Neil Bartlett was not planning to investigate inert-gas chemistry. In 1961 he was working with his graduate student Derek Lohmann in Vancouver to identify a curious salt that he'd first made while still a graduate student himself at King's College in his home town of Newcastle, England. This volatile red compound contained oxygen, platinum and fluorine – it had the chemical formula O_2PtF_6 – and Bartlett and Lohmann had come to a rather remarkable conclusion. The crystals, they said, contained oxygen molecules, O_2, in the form of positively charged ions.

That was very weird. Oxygen gas, O_2, is very good at plucking electrons from other substances – so much so that it is deemed prototypical in this regard, and any substance with a tendency to pull electrons from another material is called an oxidizing agent or 'oxidant'; the electron-depleted material that results is said to be 'oxidized'.* But if Bartlett and Lohmann were right, their strange salt was formed by the *oxidation of oxygen* – rather than sucking electrons off something else, oxygen was here having them sucked off by molecules of PtF_6, platinum hexafluoride. In other words, PtF_6 was an even more powerful oxidizing agent than oxygen itself.

Bartlett (Figure 21) was not the first person to make PtF_6 – it had been synthesized in 1958 by Bernard Weinstock and his co-workers at Argonne National Laboratory in Illinois. But the Argonne team had not studied the chemical properties of this compound. In fact, they had barely seen it, because they kept it inside metal containers, worried that it might react with normal glass. Bartlett suspects that Weinstock's team might well have made O_2PtF_6 inadvertently as air leaked into their apparatus, but the red compound would have been hidden from view.

Could PtF_6 really be oxidizing oxygen, forming a salt composed of O_2^+ ions and PtF_6^- ions? Bartlett's senior colleagues at the University of British Columbia found that hard to believe. When Bartlett submitted this claim for publication (after Lohmann had left Canada to work in Ireland), one of the paper's referees also expressed doubts. The paper got published anyway, but Bartlett was left wondering how he could prove that PtF_6 was capable of such a feat.

The amount of energy it takes to strip an electron from O_2 had already been measured. This energy is called the first ionization potential, and it is typically expressed in units of electronvolts (eV). The first ionization potential of O_2 is 12.2 eV. Now, that's a *lot* of energy: only about half that much will kick an electron out of most molecules. Bartlett realised that he might be able to prove his claim by finding another compound that is comparably hard to remove an

---

* This, in effect, tends to be the fate of materials exposed to the kind of 'oxidizing' atmosphere I discussed in the previous chapter.

Figure 21   *Neil Bartlett, who showed that inert gases are not truly inert*
(Photo courtesy of Neil Bartlett.)

electron from, and showing that PtF_6 can oxidize that too. But what to use?

'In late January or early February of 1962', Bartlett says, 'I was preparing a lecture for an inorganic class and as I was leafing through the text looking for something, the familiar plot of ionization potentials against the atomic number flicked by and I noticed that the ionization potential of xenon looked awfully like that of molecular oxygen.'

The first ionization potential of xenon is in fact 12.13 eV. So, Bartlett figured, if PtF_6 can truly ionize O_2 to make O_2^+, it should be capable of doing the same to xenon.

This is wonderfully plain reasoning. It is just like saying that if you have enough money to buy one item, you can instead buy another item with the same price. Notwithstanding the fact that no one had ever succeeded before in making compounds from the inert gases, or that everyone 'knew' it wasn't feasible, Bartlett set out to test the idea.

If there was nothing especially ingenious about Neil Bartlett's reasoning, the same was true of his experiment. That is what is so appealing about it. Having come up with the hypothesis that platinum hexafluoride might make xenon react, in March of 1962 he simply mixed them together in a flask and saw what happened. He didn't have enough money to buy much xenon, so he had to make the best of whatever he could obtain. (When he asked a colleague for xenon, Bartlett got the reply 'What do you want xenon for?' 'To oxidize it', he said, eliciting laughter that suggested he had just made a good chemical joke.)

It was late on a Friday evening when the experiment was finally ready to run, and his graduate students were at dinner, so Bartlett made the crucial test alone. 'When I broke the seal between the PtF_6 and xenon, there was an immediate reaction', he says. The reaction is rather beautiful in itself: the deep red PtF_6 gas produced a rich yellow deposit when it encountered xenon. It was obvious right away that *something* interesting had taken place.

Like Pasteur (perhaps!), Bartlett wanted to share his excitement with another person. 'So I went out into the corridor to find somebody to come and see', he says. 'There wasn't a soul in the building. So I went back to the lab.'

What *was* this yellow stuff? The best way to deduce the chemical structure and composition of a new compound was to make crystals and bounce X-rays off them (see page 161). But this didn't work for the yellow material: it wouldn't form crystals. Nonetheless, Bartlett was able to establish that the formula of the substance was $XePtF_6$: it clearly contained xenon. He wrote up his results and decided that such an unprecedented finding needed quick publication in a high-profile journal. He was sure that, now he'd published his report on O_2PtF_6, the team at Argonne who'd first made PtF_6 would soon be on the same trail, probing its striking chemical properties. 'They're going to carry out the reaction chemistry with all of these hexafluorides that they have bottled up', he thought. 'They must surely be embarrassed by the fact that for five years they had PtF_6 and they hadn't discovered that it would oxidize oxygen. They're going to turn the tables on me, so I had better get this thing out.'

So on 2 April (he intentionally avoided the first of April for fear that someone would consider the paper a hoax), he sent off his report to *Nature* in London. It was a masterpiece of concision: a mere three paragraphs sufficed to describe how to make 'the first xenon charge-transfer compound which is stable at room temperatures'. But for a month, he heard nothing. So he wrote to *Nature* saying 'Please, assure me that this paper I sent three and a half weeks ago will be published within a month. Otherwise I will withdraw it.' Still he heard nothing, and so he wrote to withdraw the paper and sent it instead to the *Proceedings of the Chemical Society*, where he'd published his work on O_2PtF_6. The xenon paper was published there in June – shortly after a card arrived from *Nature* by sea mail, simply acknowledging *receipt* of the original paper. It was an embarrassing lesson for the eminent British journal that, in science, time was now of the essence.

It took Bartlett some time to figure out how $XePtF_6$ is constituted. It is actually better written as $[XeF^+][PtF_5^-]$ – the xenon atoms do not form lone xenon ions, but bind to fluorine atoms to form XeF^+ ions. The PtF_5^- units, meanwhile, link up into long chains (Figure 22), so that the yellow material is really a kind of polymer. Later in 1962, Bartlett reacted mixtures of xenon and platinum hexafluoride containing higher proportions of the latter ingredient, and found that he could obtain products containing more fluorine, including a crystalline material $[XeF^+][PtF_6^-]$. In the meantime, the publication of his discovery stimulated others to explore xenon's new-found reactivity. Bartlett was right to anticipate how the chemists at Argonne would react: Howard Claassen, Henry Selig and John Malm in that group quickly found that xenon and fluorine could be combined directly to make xenon tetrafluoride (XeF_4), a colourless solid. That is what Yost and Kaye had tried to do in 1932, but they hadn't found the right conditions: you need to heat xenon and fluorine gases to about 400°C at six times atmospheric pressure.

Compounds of xenon are barely stable. This means not only that they are hard to make but also that they fall apart rather easily – and sometimes explosively, as Bartlett found to his cost. He decided that it should be possible to make a xenon oxide, a compound of the inert gas with oxygen. In January 1963, his graduate student P. R. Rao succeeded in making crystals of what appeared to be XeO_2, and Bartlett went along to the lab to see them. Observing standard lab practice, they were both wearing plastic face visors, but Bartlett's was rather scratched. He thought that Rao's beautiful, long crystals might contain water molecules – if so, he expected them to crumble into powder if the water were removed. So Bartlett suggested that Rao might try drying the crystals using a vacuum pump, and then he went

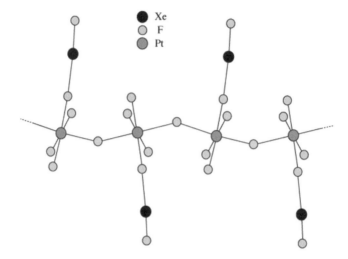

Figure 22   *The molecular structure of the first xenon compound, XePtF$_6$, is still not known for sure, because it does not form crystals. But it is expected to look something like this, with PtF$_5$ units linked into long chains. The analogous chromium compound XeCrF$_6$ certainly has this structure*

over to look at the results. 'Although I had very good eyesight', says Bartlett,

> I couldn't see the crystal surface clearly because of the visor so I put it up and just at that moment the sample, in its quartz tube, blew up. When I put my visor up, Rao put his visor up too. We were both injured in one eye, he less so than myself. I immediately lost all capability to see with my right eye because it was sliced.

They were taken to hospital, where they remained for over four weeks. By the time he got out, Bartlett discovered that someone else had already made and reported a xenon oxide, XeO$_3$. 'Things were moving so quickly', he says. A piece of glass remained stuck in his eye until it was finally removed 27 years later.

By demolishing preconceptions about the inert gases, Bartlett's discovery unleashed a flurry of activity that has now generated dozens of xenon compounds. What about the other inert gases? Krypton fluoride (KrF$_2$) was first made in 1963, and in 1962 a researcher at Argonne even reported a fluoride compound of the unstable, radioactive element radon. But the lighter inert gas elements are harder to oxidize, and it wasn't until 2000 that the first compound of argon was reported by Markku Räsänen and colleagues at the University of Helsinki in Finland. This substance, HArF, is barely stable: it exists only when cooled to within 27 degrees of absolute zero (that is, minus 246°C). No compounds of neon or helium have yet been created – they, at least, remain true to their classification as inert gases.

Embracing the alien

If it seems a little cavalier to turn the inert gases into dispassionate personalities, bear in mind that the elements do, for chemists, almost inevitably become characters in a chemical theatre. We can't help attributing distinctive qualities to them: the slow gravity of lead, the unpleasant corrosiveness of chlorine, sulphur's spicy tang and iron's ruddy humour. Nowhere is this better illustrated than in Oliver Sack's semi-autobiography *Uncle Tungsten*, where he admits what the noble gases meant to him as a solitary and isolated boy: 'I think I identified at times with the inert gases, and at other times anthropomorphized them, imagining them lonely, cut off, yearning to bond.' It is not surprising, then, that the possibility that they might form compounds became something of a personal quest for the young Sacks:

> Was bonding, bonding with other elements, absolutely impossible for them? Might not fluorine, the most active, the most outrageous of the halogens – so eager to combine that it had defeated efforts to isolate it for more than a century – might not fluorine, if given a chance, at least bond with xenon, the heaviest of the inert gases? I pored over tables of physical constants and decided that such a combination was just, in principle, possible.

Oliver was 'overjoyed' to hear about Bartlett's compound – even the loneliest of elements could, after all, find partners.

He was not alone in experiencing wonder and delight at that discovery. The physicist Freeman Dyson told Sacks that he had the same thrill at seeing xenon 'beautifully locked up in a crystal':

> One of the memorable moments of my life was when Willard Libby*
> came to Princeton with a little jar full of crystals of barium xenate.
> A stable compound, looking like common salt, but much heavier. This
> was the magic of chemistry, to see xenon trapped in a crystal.

* Libby won the 1960 Nobel prize in chemistry for inventing the technique of radiocarbon dating.

But one of the most unique accolades Neil Bartlett's discovery received was due to the impression it made on Primo Levi, whose famous book *The Periodic Table* has surely done more than any other to convey to non-chemists the idiosyncratic characters of the chemical elements. Levi begins his book with the chapter 'Argon', where he draws analogies between the inert gases – 'the Hidden' (krypton), 'the Inactive' (argon) and 'the Alien' (xenon) – and his ancestors: 'there is no doubt that they were inert in their inner spirits, inclined to disinterested speculation, witty discourses, elegant, sophisticated, and gratuitous speculation'. But are these elements really so inert, Levi asks? 'As late as 1962', he says,

> a diligent chemist after long and ingenious efforts succeeded in forcing the Alien (xenon) to combine fleetingly with extremely avid and lively fluorine, and the feat seemed so extraordinary that he was given a Nobel prize.

In fact Bartlett has not (yet) been given a Nobel prize. But the fact that Levi took it for granted that this was so speaks volumes about the impact that the first xenon compound had on chemists. It delighted and enthralled them, because it told them that in chemistry, the wonders never cease.

CHAPTER 9

Nature Rebuilt

Woodward, Vitamin B$_{12}$ and the Beauty of Economy

New Delhi, India, 1972 — Harvard chemist Robert Burns Woodward, a Nobel laureate whose ability to make molecules inspires awe throughout his profession, announces at a lecture that he and a group in Zürich, Switzerland, have finally attained the goal of a collaborative effort that began seven years earlier: to produce synthetic vitamin B$_{12}$. It is the most complex molecule chemists have ever created, and making it involved 99 scientists from 19 countries.

There is no record of whether the talk that Robert Woodward gave in India in February 1972 was like most of his others. If it was, the audience will have earned the extraordinary punchline. Woodward's talks were marathons of legendary status. If they finished within three hours, he must have been in a hurry. He would come armed with a packet of multicoloured chalks, which he used to cover every inch of blackboard space with meticulous and rather beautiful diagrams of molecules; some people remarked that the results could have been photographed and published as they stood. Woodward was serious about his chemistry, and demanded the same of his listeners. His Thursday evening seminars at Harvard would sometimes be still going strong at one o'clock in the morning. Roald Hoffmann, a theoretical chemist who collaborated with Woodward in the 1960s on work that won Hoffmann a Nobel prize, recognized that, with Woodward in particular, time ran at a different pace during lectures, and he proposed to measure it in units of milliwoodwards.

But even Woodward acknowledged that audiences need a little diversion, and he might keep them on tenterhooks by gradually consuming an entire jug of daiquiri during his talks – without any apparent effect. Perhaps his constant intake of cigarettes and coffee hardened his constitution to withstand anything he threw at it.

This was all a quintessentially Woodwardian combination: gutsy stamina, single-minded obsession and miraculous feats of memory, combined with showmanship, artistry and a dash of boastfulness. No one pretended that Woodward was an easy person to live alongside, but he was fiercely proud of his students' successes, and his challenges extracted from them the most astonishing efforts and achievements. They knew that they were being led by a man who, with a virtuosity that verged on the magical, was redefining the capabilities of their science.

The point about Woodward was not just that he guided the molecule-building fraternity – the synthetic organic chemists – to new levels of architectural complexity. He showed them at the same time that their business was truly an art, that it possessed genuine aesthetic qualities. It was largely because of Woodward that these chemists began to speak in terms of 'beauty'. In a manner reminiscent of the way Leonardo da Vinci argued for the work of the painter to be valued by Renaissance society as a genuine 'liberal art', Woodward claimed that there was 'a considerable aesthetic, cultural and artistic element in synthetic work'. In many ways, Woodward's work on the synthesis of vitamin B_{12} – a goal that he achieved in collaboration with Albert Eschenmoser's group in Zürich – represents the zenith of his achievements, and many chemists consider it to make the best case for Woodward's own position on the virtues of chemical synthesis.

The art of synthesis

The profession of the chemist is largely concerned with building molecules. That makes it an unusual kind of science; in fact, according to some definitions of the word, it is not a science at all. The kind of chemistry that, by and large, I have discussed in earlier chapters – that which attempts to understand the constitution, the building blocks, of matter – fits nicely enough into the picture of science as a discovery-based enterprise that aims to probe and to understand the world. But since the beginning of the nineteenth century, chemists began to focus not on how *nature* puts the elements together, but how *they* might do so instead. They started to recreate the substances of nature from raw ingredients, and also to produce new materials that nature has never dreamed of.

Even the ancient Egyptians knew how to synthesize substances that they could not find in the natural world. Yet chemistry before the nineteenth century was a haphazard affair, its products arrived at mostly by a combination of chance and empiricism. It was only once chemists began to perform careful and reasonably accurate analyses of the proportions of various elements in materials that they could hope to guess at how to combine nature's substances systematically to make a desired end product.

It was not like a painter mixing colours: simply having the elements in the right proportions did not guarantee that you would end up with the compound you wanted. True enough, that might sometimes suffice for making inorganic compounds such as metal salts – Nicholas Leblanc stimulated the alkali industry in 1789 when he showed how to make soda (sodium carbonate, used for bleaching and soap-making) by throwing together common salt, sulphuric acid, charcoal and chalk. And the synthesis of new, brightly coloured pigments such as copper arsenite (Scheele's Green, made by Carl Wilhelm Scheele in 1775) and cobalt aluminate (Cobalt Blue, the invention of Louis-Jacques Thénard in 1802) demonstrated that nature's chromatic bounty could be surpassed by the ingenuity of chemists. Scottish chemist William Cullen announced in 1766 that indeed one of the purposes of chemistry was to make 'artificial substances more suitable to the intention of various arts than any natural productions are'.

But it was another matter entirely for the carbon-based compounds that, being derived from living organisms, chemists called *organic*. There seemed to be an endless variety of organic compounds that could be concocted from just a handful of elements, primarily carbon, hydrogen, oxygen and nitrogen. In the 1830s and 1840s, Justus von Liebig at Giessen in Germany perfected methods of *analysing* these substances – which is to say, measuring how much of each element they contained. This supplied chemists with a *chemical formula* for the object of their studies. The compound we now call ethanol, for example, was C_2H_6O: two parts carbon to six parts hydrogen and one of oxygen. But that alone would not help you make it, for there were many compounds with very similar formulas that behaved quite differently. Indeed, dimethyl ether (as it is now called) was found to have an identical chemical formula to ethanol, yet it was clearly a different substance.

These difficulties did not, however, prevent chemists from attempting to synthesize organic molecules. It was too important a task to be forestalled by mere ignorance. Many organic compounds had key industrial and pharmaceutical uses – phenol as a disinfectant, quinine as an antimalarial drug, indigo and alizarin as dyes. These substances were extracted from natural sources – from plants and animals – which was often a laborious, costly and precarious business. How much more convenient it would be if these 'natural products' could be made in the chemical laboratory.

That was precisely what prompted the German chemist August Wilhelm Hofmann in London to attempt the synthesis of quinine in the 1840s. The drug was extracted from the bark of the cinchona tree, found first in South America in the seventeenth century and subsequently cultivated in the British and Dutch colonies in Ceylon, India and Java. Quinine was identified as the active ingredient of the cinchona extract in 1820, and its chemical formula was determined as $C_{20}H_{24}N_2O_2$. But demand for the drug, especially in colonial India, outstripped the supply from the natural source, and so synthetic quinine was urgently desired.

Hofmann set his young student William Perkin to work on the problem in the mid-1850s, and in 1856 Perkin decided that a compound called allyltoluidine, with chemical formula $C_{10}H_{13}N$, was a good starting point. Just add oxygen and remove water from two of these molecules, and the numbers add up to give the formula for quinine.

This illustrates just how much synthetic chemists at that time were working in the dark. Allyltoluidine might look from its formula alone to be almost equivalent to half a quinine molecule, but it is actually nothing of the sort. The chemists of Perkin's day had no idea how the atoms in these molecules were joined together. That is the key to organic synthesis: not which atoms a molecule contains, but how they are connected. To do this kind of synthesis in a rational and systematic way, one needs to know the *molecular structure*, the molecule's architecture. Figure 23 shows the molecular structures of allyltoluidine and quinine: here I've shown both ball-and-stick pictures depicting the atoms and the chemical bonds between them, and the pictorial formalism used by chemists, which shows the same molecules in a simpler format. In this latter scheme, the carbon atoms that make up the basic framework or backbone of the molecules are not shown explicitly – they sit at the corners and vertices of the framework. Hydrogen atoms attached to carbons are not shown in this scheme either, just to keep it simple. This way of depicting molecules requires

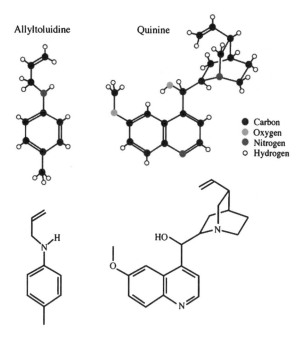

Figure 23 *Molecular structures of allyltoluidine and quinine. We can see from these structures why William Perkin was never likely to succeed in making the latter from the former*

a little practice before you can read it at a glance; but if you are not familiar with it, don't worry. All you need to recognize is that these molecules have a framework of (mostly) carbon atoms joined together in a certain way. You can see that the two molecules are really very different, so that you can't get quinine by simply welding two molecules of allyltoluidine together and jettisoning a few atoms.

Perkin did not know this, so he tried it anyway. Inevitably, he failed to make quinine; but it was his extraordinarily good fortune that after some experimentation he produced instead a compound with a beautiful purple colour, which turned out to be a very effective dye. It was the first of the so-called aniline dyes, called aniline mauve, and Perkin, who was then just 18 years old, persuaded his father and brother to join him in setting up a factory to manufacture it. Further aniline dyes followed – crimson, green, brown, black – and Perkin's discovery spawned an entire industry of dye manufacturers. The expertise in synthetic organic chemistry nurtured by this industry led to genuine successes in the synthesis of natural products towards the end of the nineteenth century, such as the production of synthetic alizarin (the colorant in madder red dye) and indigo. Dye companies no longer had to rely on the cultivation of crops to supply these compounds – which could be a haphazard affair – and the indigo plantations in the European colonies collapsed. Such was the new power of organic synthesis: it could transform the world of commerce and manufacturing.

As chemists began gradually to uncover the principles governing the shapes of organic molecules, organic synthesis acquired a new purpose beyond the production of useful materials. It became a way of checking that the molecular structure assigned to an organic compound was correct. Analysis literally means 'breaking apart': to deduce the chemical formula of a compound, chemists had to destroy it. For example, they judged the carbon content by transforming that element into carbon dioxide and measuring the quantity of gas obtained. We saw in Chapter 2 how Antoine Laviosier considered it a general principle that the composition of a substance should be deduced both by analysis and by synthesis: he verified that water was a compound of hydrogen and oxygen not just by repeating Cavendish's synthesis from these two elements but also by splitting water back into oxygen and hydrogen by passing steam over red-hot iron. For organic compounds, analysis was the easy part; but as chemists became adept at synthesis, they could use Lavoisier's principle to check their hypotheses about the constitution of molecules – by building them.

Organic synthesis, however, is hardly ever a one-step affair like the synthesis of water. The complex frameworks of the molecules must be constructed step by step: each strut, girder and bridge of the backbone is painstakingly assembled, and the frame is adorned with its various molecular accoutrements. For the purposes of structure-checking, this stepwise procedure was the whole point. At each step, the chemists knew (or at least hoped they did) precisely which links they were making, so they could keep track of the shape they were building up. If the final product then had identical properties to the compound whose structure they were probing, they could be pretty sure that they'd got the structure right. But for the chemist seeking to make a useful organic compound – a drug, say – each step of the synthesis exacts a cost. There are very few chemical reactions in organic chemistry that, however carefully conducted, convert one hundred percent of the starting material into the desired product. There are often misshapen side-products in which the chemical bonds have not formed in the manner desired; and some material is inevitably lost during the process of separating and purifying the target product, ready for the next step of the synthesis. So each step whittles away at the *yield* of the synthesis: that is, the amount of starting material converted into the final product. If each step in a ten-stage synthesis is 80% efficient (which would be pretty good going in organic chemistry), then only 10% of the starting material is transformed into the final product. This wastage pushes up the cost of the process. So as organic syntheses became increasingly complex and multi-staged, chemists were ever more pressed to keep them economically viable.

A chemical nature

Robert Woodward's first great triumph in organic synthesis was, like so much scientific innovation, prompted by the exigencies of war. During the Second World War, supplies of quinine to the USA from the Dutch East Indies were cut off. The drug was needed more than ever by troops fighting in Asia, and so there was very good reason to return to the problem that William Perkin had failed to solve.

Wartime demand for quinine was only an indirect stimulus to Woodward, however. In 1942 he was hired as a consultant to the Polaroid Corporation, the photography company based close to Harvard in Cambridge, Massachusetts. Polaroid used quinine as a light polarizer, and because the compound was in short supply, the company asked Woodward to find some alternative polarizing compounds, which he duly did. But he became lured by the synthetic challenge that quinine posed.

Things had moved on since Perkin's day – but not by very much. One of the standard approaches to the synthesis of complex natural products like this was to work backwards – to start with the product itself and convert it into simpler substances that might be easier to synthesize. Provided that one knew how to reverse the stepwise degradation of the natural product, this meant that a synthesis from scratch, using simple and readily available organic compounds, didn't have quite so far to go. In 1918 the German chemist Paul Rabe claimed to have reconstituted quinine from a degradation product called quinotoxine. In addition, in 1943 the Yugoslavian chemist Vladimir Prelog, working in Switzerland, showed how to go back and forth between quinotoxine and a still simpler compound, homomeroquinene. So if one could only get to homomeroquinene, the earlier work provided stepping stones from there to quinine.

That is what Woodward, in collaboration with William von Eggers Doering, achieved at Harvard in 1944. Their 'synthesis of quinine' was hailed by the *New York Times* as 'one of the greatest scientific achievements in a century'. But in fact Woodward and Doering may have taken too much for granted, because it isn't clear that the steps reported by Rabe really work. The problem is that quinine is an even more fiendish molecule than it appears at a glance. The molecule has no fewer than 16 different *stereoisomeric* forms, in which all the atoms are connected together in the same way but differ in their arrangements in three-dimensional space (see page 111). Only one of these stereoisomers corresponds to the antimalarial agent extracted from the cinchona tree. Rabe appeared to have no way of ensuring that the 'quinine' he could allegedly make from quinotoxine was this correct form.

This complexity arises because the quinine molecule is full of *chirality*. We saw in Chapter 6 that the bonds to a carbon atom are arranged in a tetrahedral manner, so that if each of the four attached groups is different there are two mirror-image forms called enantiomers. Such carbon atoms are said to be *stereogenic*: they generate molecular handedness. Quinine contains four stereogenic carbon atoms, each having two enantiomeric configurations; and so the total number of stereoisomers of quinine is $2 \times 2 \times 2 \times 2 = 16$ (Figure 24). (This chirality accounts for quinine's ability to polarize light.)

Many natural product molecules contain stereogenic carbons, and they are one of the biggest causes of headaches for synthetic organic chemists. When a link is forged in a molecule so as to produce a centre of chirality, it is apt to produce both of the possible enantiomers,

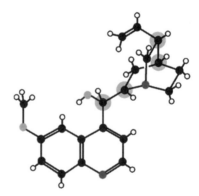

Figure 24   *The four chiral centres in the quinine molecule (circled in grey)*

whereas only one of the chiral forms lies on the path to the correct product. If the two enantiomers are made in a 50:50 mixture, half the material is then wasted (that is, the yield is at best 50%), and the two enantiomers must be painstakingly separated. Synthetic chemists try to avoid this – they are always looking out for bond-forming reactions that generate just one of the two enantiomers. In nature this is achieved by enzymes, which catalyse the formation of natural products and have an exquisite ability to direct the process towards one enantiomer or the other: they are like a glove that fits only one hand. Organic chemists now have a small arsenal of catalysts that, like enzymes, achieve some degree of chiral selectivity, but it is still a very limited toolbox. In the case of quinine, it was not until 2001 that organic chemist Gilbert Stork of Columbia University in New York succeeded in devising a complete 'from-scratch' ('total') synthesis of quinine that got all the chiral centres right.

In any event, Woodward and Doering's synthesis was never put to the test, because it would have been too expensive to produce quinine commercially that way. Besides, the war ended a year later, and so the demand fell. Nonetheless, the work established the 27-year-old Woodward's reputation as a brilliant young chemist.

That was just the start. Woodward went on to take the world of organic chemistry by storm – one complicated natural product after another yielded to his synthetic genius. In 1953 he made cholesterol, one of the components of cell membranes; in 1956 he made reserpine, which was used to treat high blood pressure and nervous disorders; and following that, lysergic acid, a pharmaceutical for circulatory, obstetric and psychological disorders (and the basis of the hallucinogenic drug LSD). But not all of his targets had pharmaceutical uses: in 1954 he made strychnine, a deadly poison. By this stage, synthetic organic chemists were starting to target 'difficult' natural products simply to flex their muscles. It so happens that several of nature's most fearsomely complicated molecules are also highly poisonous, which is why organic synthesists might appear to outsiders to have a rather unhealthy obsession with toxic substances. To a showman like Woodward, the deadly nature of strychnine may have sharpened his desire to find a way of making it: the paper in which he reported its total synthesis betrays, in Woodward's lively prose style, a certain glee in its reputation: 'Strychnine! The fearsome poisonous properties of this notorious substance attracted the attention of XVIth century Europe to the *Strychnos* species, which grow in the rain forests of the

Southeast Asian Archipelagos and the Coromandel Coast of India ...'
But even for Woodward, it wasn't all about showing off. Difficult
organic syntheses forced chemists to devise new strategies and thereby
to broaden the battery of techniques at their disposal for making other
molecules.

It was for the love of the challenge that Woodward conquered
chlorophyll in 1960 – a feat that, in the view of Albert Eschenmoser,
his one-time collaborator in Switzerland, best illustrates Woodward's
'extraordinary insight into the reactivity of complex organic molecules.' Chlorophyll, the solar cell of photosynthesis in green plants,
has at its light-harvesting heart a ring-shaped molecular assembly
called a porphyrin, studded in the centre with a magnesium atom. The
porphyrin was a new synthetic motif for Woodward, and his success
stimulated him to set his sights on another natural product with this
metal-centred ring structure: vitamin B_{12}.

A problem shared

It was, Woodward confessed, 'a monster'. Of all the vitamins, B_{12} has the
most complicated structure, and by the 1960s it was the only one that had
not been synthesized by chemists. The motivation for doing so was more
than academic. Vitamin B_{12} was used to treat pernicious anaemia, a
condition that results from a disorder in the production of red blood
cells by bone marrow. The red blood cells of people with this condition
are too few, too large, and fragile, leading to a degradation of the stomach
lining and ultimately to neurological disease. In 1926 researchers at
Harvard found that pernicious anaemia was alleviated in patients who
consumed large quantities of liver, and 22 years later researchers in
England and the USA isolated the compound responsible for this
therapeutic effect: it turned out to be vitamin B_{12}. By the 1960s the
substance was manufactured, like antibiotics, by large-scale fermentation of bacteria, and it cost about £4 per gram. Given that a typical dose
was less than a milligram, this made it a relatively cheap drug. Thus it
seemed most unlikely, given the evident complexity of the molecule, that
chemical synthesis could provide vitamin B_{12} at an even lower cost.

There *are* often good practical reasons for devising a total synthesis
of a pharmaceutical compound, even if the route is not an economically viable one. In particular, being able to build up these molecules
from scratch allows chemists to tinker with their structure, and thus to
create new molecules with related structures that might have better, or

different, activity as drugs. There is no reason, after all, to imagine that molecules synthesized in nature for one purpose cannot be improved upon when they are put to a different use in pharmacology.

Yet such possibilities were not what lay behind the attempts to synthesize vitamin B_{12}. No one much cared whether the achievement would be 'useful' or not. They pursued it for the sheer joy of the challenge, the delight at testing their skills to the limit. 'The project of a chemical synthesis of vitamin B_{12} had been taken up by its actors exclusively for purely scientific reasons, for the sake of the science of organic chemistry itself', says Eschenmoser, 'without any aspirations with regard to a possible production of the molecule – it was known that microorganisms produce it – or to create methods by which on might eventually be able to "improve" on the molecule'. (How in any case, he asks, can one 'improve' on a natural vitamin?) 'It was clear at the outset', Eschenmoser explains,

> that this vitamin represents the molecular 'Mount Everest' to the organic chemists of the second half of the twentieth century, and that its conquest would push forward the frontier to what is synthetically possible in organic chemistry. Therefore, it was a compulsory target of synthesis to the protagonist of synthetic chemistry of that time, Robert Burns Woodward.

The complexity of vitamin B_{12} became evident in 1956, when Dorothy Hodgkin at Oxford University used the technique of X-ray crystallography to deduce its molecular architecture. This method of locating the positions of atoms in crystals was introduced in the 1920s and 1930s and, as its proponents became adept enough to use it for analysing crystals of organic compounds, the role of organic synthesis for structure verification declined. Synthetic chemists gained more freedom as a result; no longer did they have to plan their schemes in terms of steps aimed at providing unambiguous indications of molecular structure. Hodgkin's solution of the structure of vitamin B_{12} was a tour-de-force that took her eight years to complete, and it was a key factor in the decision to award her the 1964 Nobel prize in chemistry. 'Vitamin B_{12} was the first complex biomolecule whose chemical structure was not elucidated by chemical degradation but by X-ray analysis,' says Eschenmoser.

Vitamin B_{12} is a most unusual organic molecule, not least because it is the only vitamin to contain a metal atom, and one of the very few biomolecules that contains cobalt. An atom of this element sits in the middle of a ring that looks, in its basic framework, very much like the porphyrin in chlorophyll: there are four small rings of five atoms joined together into a larger ring (Figure 25). But there is a subtle difference: whereas all the 5-rings of chlorophyll are linked through bridging carbon atoms, two of those in the B_{12} ring are joined directly. This structure is called a corrin ring, and each of its four 5-rings is different from the others. They are conventionally labelled A–D, as shown below. Hanging off the D ring is a side chain that joins up to a molecular group called adenosine, one of the building blocks of the DNA molecule. This loops back up under the ring so that its nitrogen atom forms a bond to the cobalt atom in the centre of the corrin ring.

So there are a lot of peculiar components to vitamin B_{12} – something that led chemists to anticipate that a total synthesis would require the blazing of new trails. What is more, the molecule has no fewer than nine chiral centres: it was a tremendous challenge to get all of these right.

When Woodward decided to embark on the synthesis of vitamin B_{12} in 1961, he found that others had already stolen a march. At the Eidgenössische Technische Hochschule (ETH) in Zürich, Eschenmoser had started working on so-called corrinoid compounds, containing the corrin ring structure, the previous year.

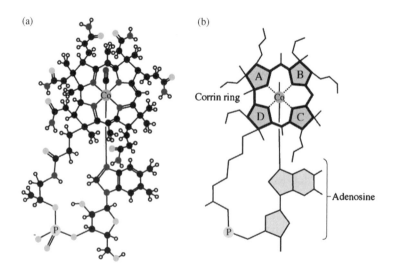

Figure 25   *Vitamin B_{12} molecule* (a), *and its key component parts* (b)

Woodward knew what was happening at ETH, because he already had strong links with the Swiss institute. He visited it in 1948 as a Swiss-American Foundation Lecturer, and developed strong ties that ultimately led Prelog at ETH to initiate the founding of the Woodward Research Institute at the labs of the chemicals company CIBA in Basle in 1963. Nonetheless, Woodward and Eschenmoser worked independently on vitamin B_{12} until 1965. They had an opportunity to compare progress the previous year at a meeting at the Royal Society in London in June, convened by Dorothy Hodgkin, which gathered together all the groups who had been drawn into the race. Among them was the British chemist John Cornforth, who later shared the Nobel prize with Prelog for their work on the chiral chemistry of organic and bio-chemical reactions. Cornforth was finding the problem hard going: he confessed that he had only 'some nice clay from which he hoped to make bricks', and eventually he abandoned his attempts to synthesize vitamin B_{12} altogether. But Eschenmoser was much further along the road, having already made the basic corrin ring structure. This drew praise and admiration at the London meeting, but was, according to Lester Smith, the scientist who first isolated red crystals of pure vitamin B_{12} in 1948, 'a long way yet from the synthesis of the complete biologically active B_{12} molecule.'* Still, Smith was optimistic that 'the complete molecule may be synthesized within a few years.'

To Eschenmoser, however, his experience so far served only to emphasize the magnitude of the task that lay ahead. Even Woodward had to concede that this was a problem he wasn't going to solve alone. He and Eschenmoser realised that their efforts up to that point had been almost perfectly complementary – they had each focused on different halves of the vitamin B_{12} molecule. The obvious strategy, then, was to combine forces and to split the effort. They decided that Eschenmoser in Europe would work on making the B and C rings – the 'east' side of the corrin ring – while Woodward in the USA would figure out the A and D rings – the 'west' side. Friendly rivalry turned into fruitful collaboration. 'We exchanged samples and information', Woodward later said, 'and visited one another and worked closely together, and were as nearly one group as we could be' (Figure 26).

---

* Some of the popular press failed to appreciate the distinction: the *New York Times*, which appeared equally hazy about who had done what, announced 'British report synthesis of vitamin B-12 molecule.'

Figure 26 *Robert Woodward (right) and Albert Eschenmoser during their
collaboration on the synthesis of vitamin B_{12}*
(Reproduced from *Nachrichten aus Chemie, Technik und Laboratorium*
(1972) (Wiley-VCH))

Ring roads

Of all the experiments in this book, this is the hardest to convey in
terms that speak about beauty. You might even ask whether a body of
work extending over two continents and 12 years can be called an
'experiment' at all. I would argue that it qualifies because it had a
single, well-defined goal and a particular strategy by which that end
was attained. As in all experiments, the art lies in this strategy. That is
the difficult part.

Synthetic chemistry has been compared to painting and to poetry, in
the sense that it represents a bringing-together of elements that must be
composed so that they work as a satisfying whole. But I think it best
resembles chess. Synthesis is a sequence of moves, each of them rather
constrained and meaningless on their own; but at several stages during
the game, one is compelled to murmur 'so that's what you're up to',
because the moves have advanced with cool logic and yet also – for a
top-class player – with flair and imagination – towards a revelatory
goal. The choice of starting reagents in synthesis is the opening game.

There is no unique way to conduct it, but that choice tells you something about the player who makes it. After the grand sweep of thrust and parry, the end game is both inexorable and modest: small gestures, carefully crafted, putting the seal on the grand strategy.

But to truly feel the beauty of chess, you have to possess some skill at the game. '[The] ability to perceive the order and reasonableness of a chess sequence', says English Grand Master Jon Levitt, 'will depend on how sophisticated the viewer is. You cannot expect to enjoy any form of chess without, to some degree, understanding it.' This echoes more or less precisely what Crystal Woodward, not a chemist but an artist (like her mother), said of her father's organic chemistry: 'For a nonchemist, trying to understand the artistic quality of Woodward's work would be like trying to read poetry in a foreign language.'

If you are not a chemist, this no doubt sounds discouraging. You may anticipate that you'll have to take my word for it that Woodward's organic synthesis was a thing of beauty. Maybe that is largely true. But let me give you some indication of the way he thought about building molecules. He once talked of synthesis as 'the design of ways to place constraints on molecular motion, with the aim of bringing about desired changes and suppressing others'. Now, this says something very striking. It is common to think about molecular structure in static terms – after all, chemists draw rigid framework-pictures of molecules, giving them a robust three-dimensional architecture. Woodward didn't see it like that. For him, molecules were fluid things. To fashion them into the shape you want, you have to grab an edge here and a strut there, and pin them in place, if only temporarily, so as to arrest the motions in that part of the structure until you have got everything properly in place. The molecules have to be cajoled, seduced, compelled to do one's bidding. At one stage in his vitamin B_{12} work, he said of his molecules that they 'behaved themselves in a way which we found most gratifying', as though they were sentient creatures with which he was grappling. In the classic portrait of him grasping a model of vitamin B_{12} (Figure 27), he seems to me to be not so much holding it up to admire the beauty of the form, as putting links in place and pulling and prodding the object to see which way he can make the framework move.

This vision of the art of synthesis guided Woodward to one of his most renowned manoeuvres, the so-called 'ring tactic', in which he would 'freeze' part of a molecule by making a ring, preventing it from any untoward gymnastics, only at the end to break the ring open and

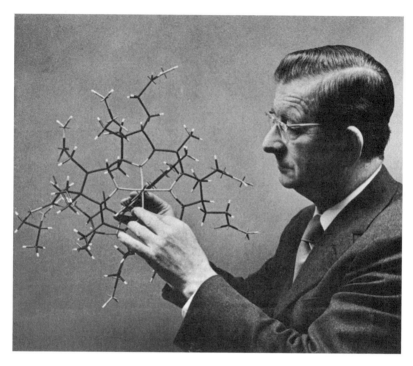

Figure 27   *Woodward, a master of synthetic organic chemistry, studies a model of the vitamin B₁₂ molecule*
(Reproduced from Harvard University Archives)

release it (Figure 28). Here again is the chess-player's mind at work: by lavishing some brief attention away from the main field of action, one keeps the whole game in order until it is time to loop back to that position and undo the holding ploy. Truly, a stitch in time.

This kind of invention and imagination is evident throughout the synthesis of B_{12}. It took something like 100 steps – an almost absurd

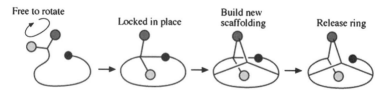

Figure 28   *How forming a temporary ring can aid organic synthesis. The ring here locks the two grey groups in the right arrangement for the subsequent steps (with the dark grey rather than the light grey group pointing up) – even though in the final molecule the ring-forming link to the black group is severed again*

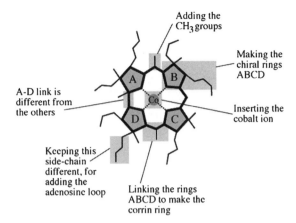

Figure 29 *The main challenges in synthesizing cobyric acid, the heart of vitamin B$_{12}$*

number for a total synthesis, even by today's standards. Before attempting to sketch out the route, let me point out some of the key problems (Figure 29). Each of the rings A to D has stereogenic carbon atoms that must be prepared in the right enantiomeric form. The rings must be joined together, with the complication that the A–D link is different from the other three. The D ring has the long side-chain linking to the adenosine group, and so it must be made somehow 'special' to allow it later to be distinguished from the others. Even putting the cobalt atom into the middle of the ring, which looks as though it should be easy, turned out to be a major hurdle. 'We found to our dismay', Woodward said, 'that cobalt was an extremely effective catalyst for the destruction of our compounds' – the corrin ring, made with immense effort, was apt to disintegrate when exposed to cobalt compounds.

The general strategy for making a complicated molecule is to break it down into little pieces that you can see how you might make, and then figure out how to put them all together. Woodward's team figured that they could make the A–D part of the corrin ring from a molecule that they called beta-corrnorsterone (Figure 30), the kind of play on words that chemists delight in. The backbone of this molecule is like that of a kind of compound called a steroid. Having made cortisone, cholesterol and reserpine, Woodward already had a box of tricks for making such molecules. The particular kind of steroid involved in this case is called a norsterone; the 'corr' prefix referred to its role as a stepping stone to the corrin ring. But, said Woodward, if you said 'corrnorsterone' with the kind of sloppy American accent he called

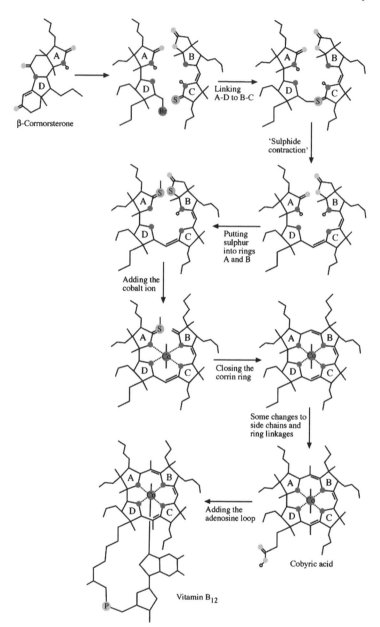

β-Cornorsterone

Linking
A-D to B-C

'Sulphide
contraction'

Putting
sulphur
into rings
A and B

Adding the
cobalt ion

Closing the
corrin ring

Some changes to
side chains and
ring linkages

Adding the
adenosine loop

Cobyric acid

Vitamin B$_{12}$

Figure 30   *An outline of the synthesis devised by Eschenmoser, Woodward and their colleagues. For clarity, I have not shown in full all the bits and pieces attached to the four rings A–D. The atoms labelled S and Br are sulphur and bromine*

Slurvian, it sounded like 'cornerstone' – which, in terms of the overall strategy, pretty much summarized the molecule's role.

While the Harvard group laid their cornerstone, Eschenmoser's team got together the B–C half of the ring (Figure 30). Then the researchers figured out how to link them together. First they made the D–C link using a method called a sulphide contraction, which had been developed by Eschenmoser's group to make their half of the molecule. After a fair bit of painful experimentation with cobalt compounds, Woodward's team found out how to put a cobalt atom into the middle of the horseshoe-shaped molecule, holding it together for forming the A–B link that completed the corrin ring (Figure 30).

A bit of theory

Robert Woodward had a favourite reaction, and he used a variant of it in one of the many steps that led to beta-corrnorsterone. It is a way of making new rings from carbon atoms, and often proved immensely useful for constructing the many-ringed backbones of steroid-like molecules. First described in 1928 by German chemists Otto Diels and Kurt Alder, the reaction involves a rather elegant concerted shunting of electrons to form two chemical bonds at once (Figure 31a). Woodward says that he heard about this reaction as soon as Diels and Alder published their report. Even though he was only ten years old at the time, the story is perfectly plausible – he was precociously advanced in his knowledge of chemistry. It captivated him ever since. But when he attempted to use a step similar to the Diels-Alder reaction

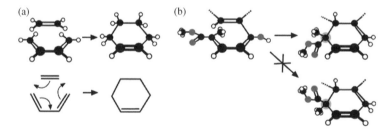

Figure 31 (a) *The Diels–Alder reaction. This chemical process can be regarded as the synchronized rearrangement of several chemical bonds.* (b) *When Woodward used a Diels-Alder-like reaction to make a part of his corrnorsterone molecule, he was puzzled by the fact that he always got just one of the two possible chiral products. Here the carbon atom at the newly formed chiral centre is highlighted in grey*

to forge links that would carry him towards beta-corrnorsterone, he found that the bonds always formed so as to give just one of two possible chiral products (Figure 31b). Moreover, if he used light rather than heat to induce the reaction, the alternative stereoisomer was formed exclusively. This was very useful – as I've indicated, it isn't usually easy for chemists to make their reactions selective about the chirality to the products. But Woodward was puzzled about what was going on. Simple-minded chemistry seemed to imply that the Diels-Alder-like reaction wouldn't have this specificity.

As we saw in Chapter 8, a chemical bond between two atoms is the result of the sharing of electrons. Crudely speaking, each atom contributes one electron to the union, and they form a pair, rather like two hands joining in a handclasp. Quantum theory says that electrons are confined to 'orbitals' – rather like the orbits of planets around the sun, except that they have some strange shapes. In the reaction that Woodward and his colleagues were trying to use, carbon atoms in the reacting molecules were linked by 'double bonds' in which the electrons occupy dumbbell-shaped orbitals (Figure 32a). To form a new bond, these dumbbells rotate through a quarter-turn. In principle it looks as though they could rotate in either direction. But it turns out that the quantum rules that determine how electrons can combine in bonds dictate that both orbitals must rotate in the same sense – both

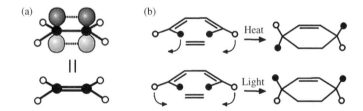

Figure 32   (a) *Double bonds are forged between carbon atoms by the overlapping of dumbbell-shaped orbitals. Here the two phases of the electron 'waves' in these orbitals are shown in dark and light grey. (b) New bonds are formed in the Diels-Alder reaction by rotation of these dumbbell orbitals. The direction of rotation is controlled by 'symmetry': the overlapping lobes of the orbitals on the two reactant molecules have to have the same phase. If the reaction produces chiral centres, as shown here, these rotations – which are described by the Woodward-Hoffmann rules – can give rise to different stereoisomers, depending on which direction they go in. The orbitals involved have different symmetries – different phases of the lobes above and below the plane of the molecule – if the reaction is induced by heat or by light, and so the products are different in the two cases*

clockwise or both anticlockwise – if the reaction is induced by heat. But because the light-induced reaction involves different arrangements of these orbitals, in that case the rotations must be in opposite senses (Figure 32b).

This was not known when Woodward puzzled over the products of his Diels-Alder-like reaction. He knew enough quantum chemistry to suspect that something of that sort was going on. But he didn't know how to turn this inkling into a proper theory. 'I very soon realized', he said, 'that I needed more help'. So he approached Roald Hoffmann, then a young chemist at Harvard who had acquired a reputation as an insightful theorist. Woodward outlined his vague ideas to Hoffmann, and said 'Can you make this respectable in more sophisticated terms?' 'He could', Woodward later recalled, 'and did.' The result was the so-called Woodward-Hoffmann rules, which described the constraints placed on the geometry of chemical processes like the Diels-Alder reaction owing to the shapes and quantum character – the 'symmetry' – of the electron orbitals involved.

The Woodward-Hoffmann rules, which these two chemists devised and refined between 1964 and 1969, show that organic synthesis, which it is tempting to regard as only a step above crude chemical bakery, in fact may depend on the acute subtleties of the quantum world. In Hoffmann's hands, Woodward's astounding intuition about how molecules behave found expression in rigorous theoretical terms. In 1972 one chemist called this discovery 'one of the few basic theoretical contributions to come along in organic chemistry in the past century'. The work won Hoffmann his Nobel prize in 1981, and Woodward would surely have shared it (making him a double laureate) if he hadn't died in 1979. Synthetic chemists who labour away making the most obscure, intricate and apparently useless molecules like to justify their efforts by claiming that they learn a lot along the way. The Woodward-Hoffmann rules show how profound such serendipitous discoveries can be.

For Woodward, this crowned his lifelong fascination with the Diels-Alder reaction. But in the B_{12} story, it was just a subplot; indeed, a subplot that went nowhere, for the reaction that led to the Woodward-Hoffmann rules turned out to be a dead end. Albert Eschenmoser, however, saw immediately that the new-found understanding of the role of 'orbital symmetry' in bond formation could be put to use in their quest. He realised that there was another way to make the corrin ring from its four components: he devised a strategy in

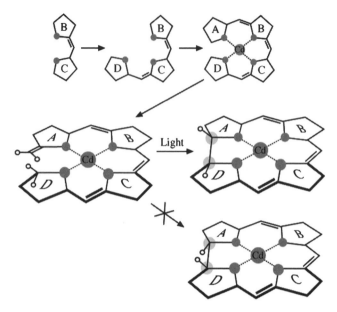

Figure 33 *Eschenmoser used the Woodward-Hoffmann rules to devise another way to make the corrin ring of vitamin B₁₂. Instead of first making the A–D and B–C units, he joined A and D separately to B–C and then forged the A–D link using a light-driven reaction. The product of this reaction was governed by the Woodward-Hoffmann rules, such that the two chiral centres created by the A–D link (highlighted in grey) had the correct geometry. Note that the metal in the centre of the ring in this case had to be cadmium (Cd) rather than cobalt – cobalt was substituted for cadmium later in the synthesis*

which his team added the A and D rings separately to their B–C unit, and then linked up A with D to complete the circle (Figure 33). The crucial point was that, by using light to stimulate the reaction that forged the A–D bond, the Woodward-Hoffmann rules dictated that the bond would have to form with just the right chiral geometry. This fiendishly clever piece of chemistry, Eschenmoser says, 'would hardly exist today, had the efforts at Harvard to synthesize the A–D half of the B_{12} molecule not stimulated (or been synchronous with) the discovery of the Woodward-Hoffmann rules. If the term 'beauty' of a reaction is ever justified in such a context, it is in this one.'

Even once the teams had made the complete corrin ring, it was still a long haul to get to vitamin B_{12}. As I indicated earlier, there was already a known route from cobyric acid to vitamin B_{12}, and so the researchers had only to take their synthesis as far as cobyric acid for the 'total synthesis' to be complete.* But that was still a very tough challenge, which the two groups didn't complete until 1972. On the brink of triumph, Eschenmoser and Woodward kept the world waiting while they perfected the synthesis. At a meeting in New Delhi in February 1972 Woodward made the first announcement of the total synthesis, but he felt that it left something to be desired. 'We had finished the synthesis', he later said, 'but the later stages were very much in need of improvement.' This low-key approach meant that it wasn't until November, when Woodward described a tidier strategy at a chemistry symposium at Wesleyan University in Middletown, Connecticut, that the world's press took notice of what he and Eschenmoser had done.

The newspapers were as euphoric as they could manage to be about this recondite art. 'Vitamin B-12 project hailed as a milestone', said the *International Herald Tribune* in January 1973. The reporters knew this was important; but they had trouble seeing precisely why.

Making it

Organic total synthesis, like chess, requires great concentration. You have to remember, in the midst of a long, many-step sequence of moves, where you are and where you are going. You have to keep an eye on what is happening at the fringes as well as in the thick of the action. There is also a lot of sheer, repetitive manual slog involved: purifying and crystallizing your products at each step, adjusting the reaction conditions to maximize the yield. Laboratories that pursue this kind of chemistry are notoriously intense, even overwhelming, places to work. In addition, because certain target molecules tend to become fashionable at any one time, there is often great competitive pressure, with several groups racing simultaneously to the latest trendy summit. Robert Woodward was the kind of person who thrives in such an environment; but it did not make life easy for those around him. There is something plaintive, even tragic, in his daughter Crystal's remark that 'He was much more often in his lab and office than he was at home, for what he wanted to be doing was chemistry; and when he

---

* Remembering his experience with quinine, Woodward was uncomfortable with relying on the earlier work of others to complete the pathway. So he and his student Mark Wuonola devised their own way of adding the loop to the D ring of cobyric acid, hooking it back up to the central cobalt atom to make vitamin B_{12} itself. But that didn't happen until 1976.

died of a heart attack at age sixty-two, it was basically because he wore himself out doing his work.'

His post-doctoral student David Dolphin made the mistake, on joining Woodward's lab, of asking about vacations, to be told by Woodward that he sometimes took Christmas Day off. But not even that, sometimes: he christened one synthetic molecule Christmasterol, because it was first crystallized in his lab on Christmas Day.

If you were in Woodward's team, he was a generous and inspiring colleague. If you were on the other side, he could be a nightmare. He was particularly prone to clashes with Robert Robinson, a British chemist who was deemed to be the master of organic synthesis in the 1920s and 1930s. They fell out in 1944, arguing over the structure of penicillin, and Robinson considered that Woodward usurped his work on strychnine. One evening in 1947 when Woodward dined with Robinson in New York, he sketched on the paper tablecloth what he believed to be the molecular structure of strychnine. According to the eminent chemist Derek Barton, once a post-doctoral student with Woodward,

> Robinson looked at it for a while and cried in great excitement 'that's rubbish, absolute rubbish!' So ever after, Woodward called it the rubbish formula and was indeed quite surprised to see, a year later in *Nature*, that this was also the formula deduced eventually by Robinson. Woodward should have sent the table cloth for publication!

Barton also tells a story (probably apocryphal) of how the two men happened to meet on the platform of the railway station at Oxford one morning in 1951. Robinson asked Woodward what he was currently working on, and hearing that the answer was 'cholesterol', he shouted 'Why do you always steal my research topics?' and hit Woodward with his umbrella.

Yet what Woodward constructed in his hot-house lab was a new era of organic synthesis. It seemed to those who watched him scale every new peak, each one higher than the last, that Woodward had proved chemistry had no limits. After the synthesis of vitamin B_{12}, Eschenmoser says 'many chemists were led to believe that Woodward had shown in principle that any natural product, be it ever so complex, would be amenable to synthesis, given only a great enough investment of time and resources'. Thanks to Woodward, it seemed that chemists could stand alongside nature and hold their heads up high.

CHAPTER 10

Plato's Molecules

Paquette's Dodecahedrane and the Beauty of Design

Columbus, Ohio, USA, 1980—A team of organic chemists at Ohio State University, led by Leo Paquette, create a molecule of near-Platonic perfection that has tantalized and eluded chemists for at least two decades. This is dodecahedrane – or rather, at this stage a molecule closely related to it. It contains a dodecahedron made from 20 carbon atoms linked into roughly spherical structure, and it demonstrates chemistry at its most sculptural: a molecular object valued purely for its beautiful symmetry.

Scientists tend to espouse a concept of beauty that is at once both esoteric and rather formal. They talk a lot – surprisingly so, perhaps – about 'beauty', but what do they mean by it? 'Physical laws should have mathematical beauty', said the British physicist Paul Dirac, but how many people will find the beauty in his Nobel-winning equation, which reads like this:

$$\left[\gamma^{\mu} \left(i \frac{\partial}{\partial x^{\mu}} - e A_{\mu}(x) \right) + m \right] \psi(x) = 0$$

Einstein said 'the only physical theories that we are willing to accept are the beautiful ones'. But is relativity beautiful? Is quantum theory beautiful?

Well, they were beautiful to people like Einstein and Dirac, and one of the prime reasons why that was so is that deeply embedded within these theories is the concept of *symmetry*. Many scientists, particularly

physicists, acquire a kind of X-ray vision for symmetry, enabling them
to see this mathematical property where others see just a jumble of
symbols or a scattering of data points. They, like the Platonist scholars
of the Middle Ages, equate the symmetrical and the orderly with the
'good'. While they commonly mistake this aesthetic response for an
artistic one, it is in fact much closer to a theological, almost a moral
judgement: scientists derive a strong sense of satisfaction and 'rightness'
from symmetry.

Chemists love symmetry too, and for much the same reasons. But
they are a more visual community, and their aesthetic sense is considerably easier for the non-scientist to appreciate. Chemists are attracted
to symmetrical objects, and none more than the so-called Platonic solids
(Figure 34), the three-dimensional shapes that Plato, in his *Timaeus*,
called 'the most beautiful bodies in the whole realm of bodies'.
Mathematician Hermann Weyl, in his classic book *Symmetry* (1952),
considered that Plato's identification of these bodies was 'one of the
most beautiful and singular discoveries made in the whole history of
mathematics'.

For Plato, who was strongly influenced by Pythagoras's sense of the
divinity of geometry and number, these five solids were in some way
the fundamental constituents of all things. He equated four of them with
the elements identified by Empedocles: the tetrahedron was fire, the cube
earth, the octahedron air and the icosahedron water. The fifth solid, the
dodecahedron, became identified with the 'fifth essence', the aether that
was thought to fill all the heavens. And so in a sense the dodecahedron

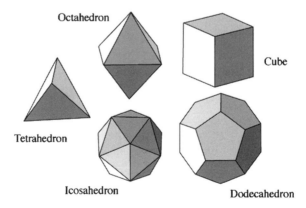

Figure 34   *The five Platonic solids*

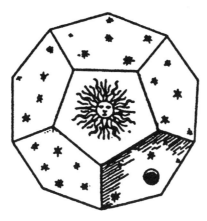

Figure 35 *The dodecahedron symbolizes the universe in Kepler's* Harmonices mundi

came to represent the universe, as depicted in Johannes Kepler's *Harmonices mundi* in the early seventeenth century (Figure 35).

Plato blended the Pythagorean notion of geometrical beauty with the atomism of Democritus, who preceded him by just one generation. Thus, Plato's elemental 'atoms' were deemed to have geometric shapes corresponding to their respective element, which seemed to account for some of their properties. Fire penetrates and stings because of the sharp points of the tetrahedral atoms; water flows because its ico-sahedral atoms are so nearly rounded; the stability of earth is explained by the square faces and thus the stable packing of its cubic atoms, while the triangular faces shared by the other three elements make them interconvertible.

It is no surprise that a Neoplatonist like Kepler should have per-petuated these geometric ideas about the constitution of matter. He extended them to develop a theory of the planetary orbits, in which each of the six planets known in his time follow circular orbits whose distance from the sun* was determined by their need to fit neatly into a 'cage' shaped like a Platonic solid. There was no real scientific motivation for this theory; it was essentially an expression of Kepler's mystical beliefs. But there seemed to be plenty of evidence that nature had a fundamentally geometric aspect. It was Kepler, after all, who gave the first geometrical explanation for the symmetry of crystals, when he argued in 1611 that the sixfold symmetry of snowflakes stems

---

* Most astronomers by that time acknowledged Copernicus's heliocentric universe, even though the Roman church continued to resist the idea.

from the regular packing together of the 'globules' of water from which they are comprised.

Friedrich August Kekulé's discovery of the benzene ring, and Jacobus van't Hoff's realization that carbon chemistry is dominated by a tetrahedral arrangement of atoms (page 111), indicated to nineteenth-century chemists that theirs was a deeply geometrical science. At much the same time, microscopic studies of tiny marine organisms called radiolarians revealed their astonishingly Platonic character: they have mineralized cages (exoskeletons) with polyhedral shapes, rather like Plato's regular bodies (Figure 36a). The icosahedron was found in the mid-twentieth century to be echoed at a still smaller scale in the shapes of viruses (Figure 36b).

Against this context it is perhaps inevitable that chemists should wonder if they might construct molecules shaped like the 'beautiful' Platonic polyhedra. Could they arrange atoms into frameworks that reflect the symmetry of Plato's 'atoms'?

Basket chemistry

The concept of molecular shape that stemmed from the work of Pasteur and van't Hoff forced chemists to think in three dimensions. Nonetheless, the notion of polyhedral molecules requires a further leap of imagination, for here are objects that *enclose space*, that have *volume* and which are bent and curved and that, as a result, might be strained like the shaft of a longbow. It is one thing to think of molecular chains in which the bonds wiggle this way and that, and even of rings in which the ends of a chain are joined up; but these three-dimensional molecules are truly a form of atomic architecture.

Chemists were introduced to such ideas when in 1933 two Czechoslovakian chemists named S. Landa and V. Machacek discovered that among the complex mixture of hydrocarbon molecules in petroleum there is one that possesses such a cage-like structure. It is called adamantane (Figure 37): the name comes from the Greek *adamas*, meaning 'diamond', for the molecular framework of adamantane is like a tiny piece cut out of the carbon-atom lattice of a diamond crystal. It has the same formal symmetry as the tetrahedron: it might not *look* like a tetrahedron, but all the rotations and mirror reflections that leave the molecule looking unchanged are the same as those of a tetrahedron. There is now known to be a whole host of these 'diamondoid' cage molecules in crude oil.

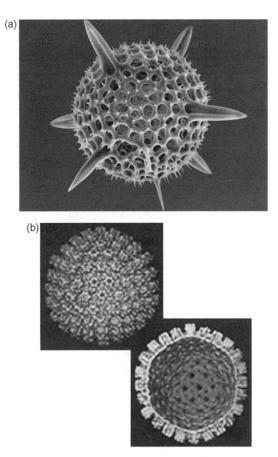

Figure 36 *Platonism in nature.* (a) *Polyhedral exoskeletons of radiolarians.* (b) *Many viruses have icosahedral shapes* (Part (a) © D. Breger, Drexel University, part (b) courtesy of the Trustees of Dartmouth College)

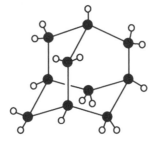

Figure 37 *The adamantane molecule*

If nature can make such carbon-framework cages, surely synthetic organic chemists could find a way to do it? The most appealing targets were the hydrocarbon molecules that represented the Platonic solids themselves, in which a carbon atom sits at each vertex of the body, and the edges are represented by chemical bonds between them. But not all of the solids can be constructed this way: some are simply not chemically feasible. Carbon atoms can form no more than four chemical bonds*, and we have seen that these are oriented around the carbon atom such that they point (more or less) to the four corners of a tetrahedron.

The former consideration rules out the icosahedron, because each of its vertices is connected to five others – that's too many bonds for a carbon atom. The octahedron, meanwhile, has only four links to each vertex, but they are in the wrong arrangement to be sustained by a carbon atom – this would require such a distortion of the usual tetrahedral geometry of the bonds that the molecule would not tolerate it. The bonds would be too strained, and they'd just pop open.

The three other Platonic solids, however – the tetrahedron, cube and dodecahedron – each have three edges per vertex. So in principle they could all be made from carbon frameworks, with the fourth bond that each carbon atom requires pointing away from the cage like a pro-truding spike. These 'capping' bonds may be supplied by hydrogen atoms, as they are in adamantane. Every carbon atom will then have a single hydrogen atom attached to it, and so the chemical formulae for these 'tetrahedrane', 'cubane' and 'dodecahedrane' hydrocarbon mole-cules are C_4H_4, C_8H_8 and $C_{20}H_{20}$ (Figure 38).

That sounds good in theory; but can such molecules exist? For a carbon atom with a perfectly tetrahedral arrangement of bonds, as in methane (CH_4), the angle between each adjacent pair of bonds is approximately 109.5° (Figure 39a). In tetrahedrane, the angle between adjacent edges is just 60° (Figure 39b). In other words, each carbon's 'tetrahedron' must be severely bent. It is not obvious that the molecule could survive such a straining of the normal bond angles – and indeed, no one has, yet, made tetrahedrane. But the basic tetrahedral carbon framework *can* be created if, instead of capping each carbon with a hydrogen atom, they are capped with a much bigger, bulkier chemical group. This surrounds the carbon tetrahedron with a kind of shell that holds it in place: a phenomenon dubbed the 'corset' effect. In this way,

* There are exceptions to this, but they are extremely unusual and rare. Molecules containing carbon atoms with more than four bonds are called 'hypervalent', and they are 'non-classical' in the sense that the bonds are not formed by the usual mechanism of simple electron-pairing.

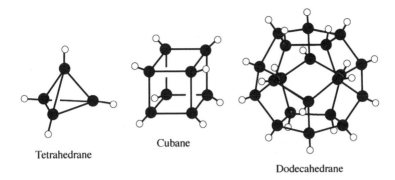

Cubane

Tetrahedrane

Dodecahedrane

Figure 38   *Tetrahedrane, cubane and dodecahedrane*

Günther Maier of the University of Marburg and his co-workers succeeded in making the first tetrahedrane *derivative* in 1978: a molecule with a kind of hydrocarbon 'bush' attached to each vertex (Figure 40).

Cubane is less distorted: the bond angle at each vertex is 90° (Figure 39c). That is not so very far from the ideal angle of 109.5°, and so this simple hydrocarbon molecule is relatively stable. Cubane was first made in 1964 by Philip Eaton and Thomas Cole at the University of Chicago. The crucial step in their synthesis was one in which they persuaded a 'polycyclic' molecule, containing various carbon rings, to fold up into a kind of 'protocube' (Figure 41), using the favourite ring-forming reaction of Robert Woodward, the Diels-Alder reaction.

Cubane is an unusual hydrocarbon: it forms colourless crystals that don't melt until they are heated to 130°C (compare that with octane,

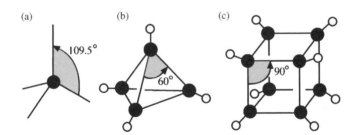

Figure 39   *The ideal bond angle for a 'tetrahedral' carbon atom is 109.5° (a). But in tetrahedrane three of these angles at each carbon atom would be just 60°, creating a lot of strain (b). In cubane the bond angles are a little wider (90°), and the molecule is consequently more stable (c)*

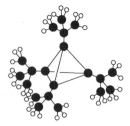

Figure 40 *Günther Maier's t-butyl tetrahedrane*

a hydrocarbon that also has eight carbon atoms but joined in a straight chain – it melts at minus 56°C). The molecule itself is very resistant to being damaged by heat: only above 220°C does it start to fall apart into other hydrocarbons. Its synthesis blazed a trail to various other highly symmetrical 'cage' molecules, such as triprismane (1973) and pentaprismane (1981) (Figure 42).

Philip Eaton's successful synthesis of cubane led him inevitably to set his sights on the third of the possible Platonic hydrocarbons: dodecahedrane. But he wasn't the first to do that. Robert Woodward himself, the maestro of carbon-molecule architecture, first began thinking about dodecahedrane in the early 1960s. In the same year that Eaton made cubane, Woodward described how to construct a bowl-shaped hydrocarbon called triquinacene, in which three of the five-membered carbon rings that constitute the faces of dodecahedrane are fixed together along their edges (Figure 43a). It looks at first glance as

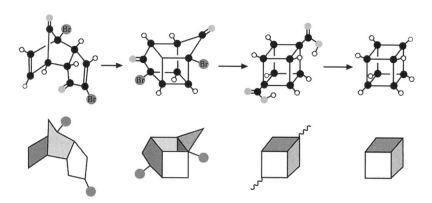

Figure 41 *The steps that Philip Eaton used to fold up a polycyclic molecule into the basic structure of cubane*

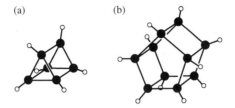

Figure 42   *The hydrocarbons triprismane* (a) *and pentaprismane* (b)

though there is still a long way to travel from this small fragment to the fully closed shell of dodecahedrane; but in fact one can imagine making that journey in a single leap. If two triquinacene molecules could be brought together in the right orientation, Woodward realised, their rims might be linked up to form the entire cage (Figure 43b).

But that was a lot to ask, and Woodward never figured out how to bring the two molecules into the right alignment. Eaton decided to try another strategy, which he hoped might leave less to chance. He built a more substantial part of the dodecahedral framework: a bowl-shaped unit that comprised half of the entire cage. Like Woodward, he hoped it might be possible to complete the remaining synthesis in a single master-stroke in which a lone pentagonal ring would form a roof over the bowl, linking up with the five peaks around its crenellated rim (Figure 44). Eaton christened the bowl-shaped molecule peristylane, after the Greek word *peristylon*, signifying a group of columns arranged around an open space and designed to support a roof. Eaton's group made peristylane in 1976, but again the challenge of positioning its pentagonal roof, and creating all the links, proved too difficult.

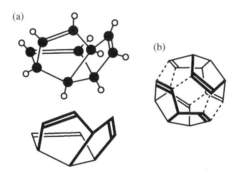

Figure 43   (a) *Triquinacene.* (b) *If two triquinacene molecules could be stitched together in the right way, dodecahedrane would be produced*

(a)   (b)

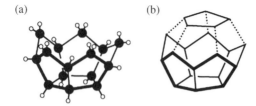

Figure 44 *Eaton's peristylane* (a) *and how the 'roof' was intended to be fitted to make dodecahedrane* (b)

Doing it the hard way

In the mid-1970s, organic chemist Leo Paquette at Ohio State University (Figure 45) decided that Woodward's notion of uniting two triquinacene bowls might be more feasible if they were given a helping hand by first tethering them together. A bond linking one rim to the other might help them find the alignment needed to close the shell. In 1975 Paquette and his co-workers devised a way of making these linked bowls, which, in recognition of their clam-like shape, they called bivalvane (Figure 46). But would bivalvane close its jaws? It would not.

All of these strategies for making dodecahedrane – via triquinacene, peristylane and bivalvane – rely on the hope that two fragments of the cage molecule might somehow be zipped together through the concerted formation of several struts in the polyhedral framework. They are, in the parlance of organic chemists, *convergent* strategies, dependent on the coming-together of several parts of the structure at once. While exploring the bivalvane approach, Paquette acknowledged that the solution might not be that simple. It might, he conceded, be necessary after all to build the carbon shell by painstakingly fashioning one strut at a time.

Indeed, that was how, in the end, Leo Paquette and his colleagues made dodecahedrane. It might sound like a disappointingly literal-minded, plodding solution. Yet I feel that, on the contrary, the synthesis of dodecahedrane is unusually clear and satisfying, for two reasons. First, it is very easy to see the logic of the synthesis. Some complex molecules are made by routes that seem circuitous or meandering, occasionally appearing to take two steps back for every step forward, while others (Woodward was a master of this) seem for much of the time to be progressing towards a molecule quite different from the intended target, only to reveal the end product through a final, surprise twist. But we can watch Paquette's dodecahedrane taking shape by a

Figure 45 *Leo Paquette, architect of the first synthesis of dodecahedrane* (Photo courtesy of Leo Paquette.)

Figure 46 *Bivalvane, a failed precursor to dodecahedrane*

methodical and transparent sequence, at each stage of which a new pentagon is inserted into the framework.

Second, the synthesis preserves the element that motivated it in the first place: symmetry. At each stage of the game, the scaffold retains an appealing degree of symmetry. For most of the time, the left-hand side of the molecule looks just like the right: you could stick a mirror through the middle, and the reflection of either half would reconstitute the whole.

The total synthesis that Paquette and his colleagues reported in *Science* in February 1981 contained 20 steps. While only an organic chemist could readily make sense of the cryptograms in which these steps were encoded, I hope you can get a sense of the symmetrical, 'face-by-face' path if I pick out some of the key framework structures that were involved *en route* to the final Platonic form (Figure 47).

It was not dodecahedrane itself that the Ohio State chemists made, however. Their synthesis left two methyl (CH_3) groups dangling from the dodecahedral framework: this was dimethyl dodecahedrane. Like cubane, it formed colourless crystals: a sign that the compact, geometrical molecules stack together in an orderly arrangement, rather like giant atoms. The researchers were unable to melt these crystals at all: when they heated them up to over 400°C, the solid compound started to turn brownish, a sign that it was decomposing in the heat to form other, tarry hydrocarbons.

It was not until the following year that Paquette's team figured out how to eliminate the dangling methyl groups, delivering them genuine

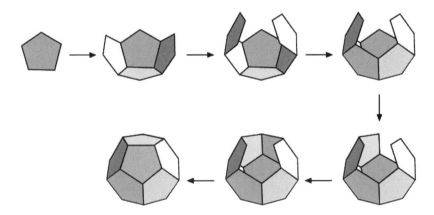

Figure 47   *Key shapes involved in Paquette's synthesis of the dodecahedrane cage*

dodecahedrane, $C_{20}H_{20}$. By then, the world had already been captivated by this most elegant of the Platonic solids. 'Plato's perfect solid produced', said one newspaper headline. But what good is it, others wanted to know? Who cares, replied the chemists – it's beautiful.

Chemistry in the round

There is now more than one way to make dodecahedrane. One of the most attractive schemes, as well as one of the most productive, was devised during the 1980s and 1990s by Horst Prinzbach of the University of Freiburg and his co-workers. Its appeal lies in the intriguing shape of the key 'intermediate' molecule along the synthetic pathway: a shape that put the German researchers in mind of a Chinese pagoda. So they called it pagodane (Figure 48). Prinzbach's graduate student Wolf-Dieter Fessner played a particularly key role in figuring out how to make this molecule from a plentiful starting material known as isodrin, which was used industrially to make an insecticide called endrin.

Pagodane has in fact precisely the same chemical formula as dodecahedrane: $C_{20}H_{20}$. The two molecules are isomers. So having deduced, in 1983, how to make pagodane, Prinzbach and Fessner suspected that they might be able to persuade the molecule to rearrange itself into dodecahedrane by simply shifting around some chemical bonds and hydrogen atoms. And so they could: warming pagodane to 300°C in the presence of a catalyst of palladium metal and carbon brought about this spontaneous 'isomerization' reaction. But a lot of other compounds were formed in the process too: only about 2% of the product consisted of dodecahedrane. So it wasn't a very efficient way to make it. Instead, Prinzbach spent several years working out how to perform the conversion systematically by taking out some of the 'struts' in the pagodane framework and putting in others. The key steps were first to open up the 'waist' of pagodane, and

Pagodane   Dodecahedrane

Figure 48 *How to make dodecahedrane via the pagodane molecule*

then to join up the ends (Figure 48). By the late 1990s, Prinzbach could carry out this conversion efficiently enough to make about a gram or so of dodecahedrane, which was considerably more than Paquette's initial synthesis produced.

But what good *is* it, supposing that you could manufacture buckets of the stuff? After Paquette reported his synthesis of dodecahedrane, some speculated that it might act as an antiviral agent, particularly against influenza. Or maybe it could be used to entrap drug molecules inside the cage and ferry them into cells, the compact spherical molecule punching its way through cell walls. (That, however, never looked feasible – there was barely enough room inside the carbon cage to fit even a single sodium ion, let alone a drug molecule.) Paquette wondered whether the spherical shape might enable the molecules to roll over one another like ball bearings, making the compound a good lubricating agent.

None of these ideas ever amounted to anything. But dodecahedrane's significance for chemistry was more far-reaching. If carbon could be fashioned into a cage this complex, could it form even larger shell-like molecules? Dodecahedrane was still fresh in everyone's minds when, in 1985, a team of chemists working at Rice University in Houston, Texas, claimed that they had created carbon cages consisting of no fewer than 60 atoms. The extraordinary – some felt unbelievable – aspect of this claim was that it did not stem from the kind of sophisticated and labour-intensive synthesis that Paquette had used. Instead, the Rice researchers simply evaporated graphite – pure carbon – using an electrical discharge. They found sixty-carbon (C_{60}) molecules in the soot that condensed from the carbon vapour.

According to its discoverers – Harry Kroto, a British chemist from the University of Sussex, along with Rice scientists Richard Smalley, Bob Curl and their graduate students – C_{60} was a carbon molecule shaped like a soccer ball: a sphere made from twelve pentagonal carbon rings and 20 carbon hexagons. Unlike dodecahedrane and cubane, it was not a hydrocarbon: there were no hydrogen atoms capping the carbon atoms at each vertex of the framework. Instead, carbon's propensity to form four bonds was satisfied by the presence of double bonds within the carbon rings (Figure 49a). This made the shell of C_{60} akin to graphite, in which hexagons of carbon atoms are joined edge to edge in vast, flat sheets (Figure 49b).

It was hard to imagine that carbon atoms could spontaneously assemble from a vapour into soccerball molecules, given how much

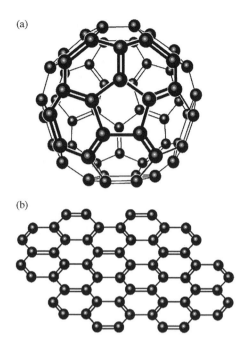

Figure 49 *The structure of the C_{60} molecule* (a) *is rather like one of the carbon sheets of graphite* (b) *curled into a ball. The curvature is produced by the presence of pentagonal rings in the framework*

chemical trickery had been required to make the smaller dodecahedrane. But five years after Kroto and colleagues presented their tantalizing evidence for C_{60}, named buckminsterfullerene in honour of the molecule's supposed resemblance to architect Richard Buckminster Fuller's geodesic domes, their claim was vindicated. Researchers in Arizona and Heidelberg figured out how to make C_{60} in large enough quantities to form crystals, allowing them to study the molecular structure using X-ray crystallography. Sure enough, C_{60} proved to be a hollow, essentially spherical carbon molecule, a polyhedron technically called a truncated icosahedron.

Some of the applications initially proposed for C_{60} were strikingly similar to those of dodecahedrane: a molecular lubricant, a drug carrier (various atoms and ions, if not drug molecules, can indeed be encapsulated within the spherical shell), an antiviral agent (it shows some activity against the AIDS virus HIV). It turned out also to have remarkable electronic properties: C_{60}, like graphite, conducts electricity, and if it is laced with alkali metals and cooled to very low temperatures it becomes a superconductor, conducting electrical current with no resistance at all.

Chemists began finding ways to stick molecular groups onto the surface of C_{60}, prompting Harry Kroto to talk about a new era of 'round chemistry'. C_{60} proved to be just one of a whole family of carbon-cage molecules made from pentagons and hexagons, dubbed fullerenes, which were also produced by evaporating graphite. Some fullerenes are giant cages, as big as C_{240} or more; others are smaller molecules, even right down to C_{20}, a perfect dodecahedron made from twelve pentagons – dodecahedrane with all its hydrogen atoms removed. And in 1991 Japanese scientist Sumio Iijima reported that graphite-like sheets of carbon could be fashioned not only into balls but into tubes: long cylinders of carbon just a few nanometres wide, called carbon nanotubes. These have been used to make electrically conducting molecular wires, molecular-scale electronic devices such as transistors, and tiny filaments with a strength and stiffness even greater than that of conventional carbon fibres.

Kroto, Smalley and Curl were awarded the 1996 Nobel prize in chemistry for their discovery of the fullerenes. But arguably it was dodecahedrane that first got the ball rolling, by showing that organic chemistry is literally multi-faceted and that carbon can be sculpted and moulded in ways that would have delighted the ancient philosophers.

CODA

Chemical Aesthetics

For Proust it was the aroma of a madeline; for me there is a particular, bittersweet benzoic smell that transports me instantly to another place and time. I am back in the organic chemistry labs of the Dyson Perrins building in Oxford, desperately scratching the inside of a beaker with a glass rod in the hope that it will seed the growth of crystals from out of a clear, warm liquid. Stepping into a chemistry laboratory is a sensually intense experience, and every time I do it, it triggers this remembrance of times past.

Smells are, of course, uniquely evocative. And there's no doubt that chemistry has a plentiful supply of the kinds of stinks that define its common caricature. Once you've taken a strong sniff, you will never forget the rotten-egg pong of hydrogen sulphide, the urinary tang of ammonia or the eye-watering sharpness of sulphur dioxide. But these are just a small part of chemistry's aromatic palette, which encompasses the fruitiness of esters, the cloying sweetness of acetone, the electric piquancy of ozone – all of them smells known in childhood (mine, at least), from pear drops and model-aircraft glues and the transformers that power model railways. With such an accumulation of reminiscences, the lab can be a disorientating place, as poignant and unsettling as a visit to one's old school.

The nose is a useful instrument for the chemist. Often smell alone will provide the vital clue about what has happened in the test tube. It was the smell of ozone that told Neil Bartlett he had found in platinum hexafluoride an oxidizing agent powerful enough to oxidize oxygen itself (Chapter 8). Chemical ecologist Tom Eisner has learnt to recognize the chemical constituents of complex mixtures exuded by insects: a skill surely honed in his father's perfumery. 'I'm essentially a nose with a human being attached', Eisner says. Primo Levi explained that he took measures to 'educate his nose'; chemistry Nobel laureate Robert Mulliken's fondness for the odours of chemicals led him to admit 'I feel I am somewhat of a dog.'

But the aromas of chemicals have a significance that goes beyond bare utility. Smell is a central component of chemistry's *aesthetic*. Equally important, equally critical to the chemists' emotional connection to their subject, is the role of colour. 'What attracted me [to chemistry]', says Bartlett, 'was the fact that one could make such beautiful materials both in color and in crystallinity. One could get these beautiful, royal blue crystals and make things.' That's how it was for Oliver Sacks too:

> My father had his surgery in the house, with all sorts of medicines, lotions, and elixirs in the dispensary – it looked like an old-fashioned chemist's shop in miniature – and a small lab with a spirit lamp, test tubes, and reagents for testing patients' urine, like the bright-blue Fehling's solution, which turned yellow when there was sugar in the urine. There were potions and cordials in cherry red and golden yellow, and colourful liniments like gentian violet and malachite green.

I savour these colours in the study where I write. Here I keep bottles and jars of chrome yellow, cadmium red, ultramarine, viridian: all of them artists' pigments bought from an art supplier, but they are of course all chemical compounds too, and my pigment cabinet might just as well be the storage shelves of a nineteenth-century chemical laboratory, in the days before these beautiful powders were hidden away within opaque plastic containers. I thrill even to see those ranks of white polypropylene jars in the modern laboratory, each with a label that gives symbolic clues to the chromatic glories within.

Colour, too, is diagnostic for the chemist. A colour change signals the completion of a reaction, just as it did for the alchemists (who regarded colour as an altogether more profound, even a spiritual, property of their materials). It was a beautiful purple colour that told William Perkin he had made a potential dyestuff – the first aniline dye, the compound that launched the modern chemical industry – in 1856 (page 155). Colour is a subtle badge of constitution: it is how chemists can tell how many electrons a manganese ion, say, has lost or gained. An experiment in chemistry can be beautiful not because of conceptual elegance or demonstrative power but merely because it is lovely to watch.

These sensual characteristics of chemical work are not merely incidental – they can provide a direct source of inspiration to researchers. They make the laboratory an alluring and exciting place to be in. Synthetic chemist Robert Woodward (Chapter 9) was the sort of scientist one might imagine scoffing at the nebulous theorizing of the art world, but he had no hesitation in confessing how the aesthetics of chemistry spoke to his soul:

> It is the *sensuous* elements which play so large a role in my attraction to chemistry. I love crystals, the beauty of their form – and their formation; liquids, dormant, distilling, sloshing (!), swirling; the fumes; the odors – good and bad; the rainbow of colors; the gleaming vessels of every size, shape, and purpose. Much as I might *think* about chemistry, it would not exist for me without these physical, visual, tangible, sensuous things.

Ah yes, the gleaming vessels. Chemists even love those too, as I am sure Ernest Rutherford loved the exquisite glass capillary that allowed him to deduce the nature of the alpha particle (Chapter 4). The Museum of the History of Science in Florence is filled with objects that are far more beautiful than they have any right to be (Figure 50): these are exhibits of sublimated artistic joy, made by glassblowers and metal

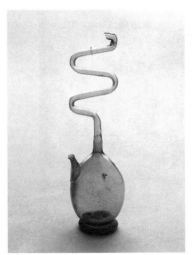 

Figure 50 *Some of the glassware made and used by chemists in former times is far too beautiful to be purely utilitarian*
(Photos: Institute and Museum of the History of Science, Florence.)

smiths and instrument makers (and scientists) who could not help but fashion something designed to be more than just functional. Even the stock apparatus of the chemist has, to my mind, a magnetic, evocative power: here are the receptacles in which matter is separated, purified, concentrated, created. That is recognized, I think, in New York artist Eve Andrée Laramée's sublime installation entitled *Apparatus for the Distillation of Vague Intuitions* (Figure 51), a scaffold populated by glass flasks and tubes and cylinders, some half-filled with mysterious residues and bearing the faint legends of etched inscriptions that hint at the way the chemist operates on the borders of what is predictable: 'matter of chance', 'leap in the dark', 'secret process of evaporation'.

Yet, despite all this, the aesthetic of chemistry (and I know of no other science that has such an explicit aesthetic element to it) has received rather little serious consideration. It is treated as though it were a happy irrelevance, a stimulus for adolescent enthusiasm or idle reverie. Chemists are, it seems, at risk of forgetting the relationship that their forebears had with the materials *they* used, which were always more than they seemed. Just as the alchemist saw a symbolic significance in the process of separating vapours ('spirits' and 'quintessences') from solids and liquids, so artists and artisans believed that the quality of their work was as dependent on the quality of the materials as on the inspiration that guided the hand. Moreover,

Figure 51   *Eve Andrée Laramée's* Apparatus for the Distillation of Vague Intuitions *(1994–98)*

for the medieval artists who built the great Romanesque and Gothic cathedrals, the material was a gateway to the transcendental: 'The dull mind rises to truth through that which is material', said Abbot Suger, the twelfth-century mastermind of the prototypical Gothic church at St Denis, near Paris.

This is not to say that chemists stand to benefit by developing some mystical relationship with zinc sulphate and nitrobenzene. Rather, it might simply be interesting to direct conscious attention to the impulses that, more or less subliminally, in fact guide some of what they do. I have said a little in Chapter 9 about Woodward's universally acknowledged artistry in chemical synthesis, which some of his peers regarded as having an almost poetic quality. The poet possesses not only the technical competence needed to put the words in the 'right' order, but an emotional and aesthetic response to words that enables him or her to find an order that goes beyond syntactic accuracy and reaches into the realm of true creativity. For there is of course no 'right' order of words, any more than there is a right way to put a complicated molecule together: one poet's choice will be different from another's, and the result will depend on how the poet relates to the words selected. By the same token, Woodward's daughter Crystal said that 'the delight and aesthetic pleasure [he took from his products] contributed to his skill'.

More generally, chemists are frequently motivated to make a molecule purely for the joy of it, as I discuss in Chapter 10. But what, really, is this motive, this sense of what is 'beautiful' in a molecule? Does it bear any comparison with what we find beautiful in art? Joachim Schummer has argued persuasively that the criterion of 'beauty' commonly applied by chemists to the molecules they make has little connection with the aesthetic measures that have traditionally been applied to art. The chemist tends to find beauty in a high degree of symmetry, in a Platonic orderliness and regularity. This was indeed the notion advanced in Plato's *Timaeus*, which informed the aesthetic sensibilities of the early Middle Ages through the philosophy of Neoplatonism. But it was never really a part of any *artistic* theory: Plato made symmetry the measure of beauty in *nature*, whereas he had a rather low opinion of the plastic arts such as painting and sculpture, distrusting their 'deceptiveness'. The symmetry of the Gothic cathedrals stems from a basically religious philosophy, from a worldview that equates order with morality. To our eyes today, much of the real beauty of these edifices stems from other features: the glorious use of light (although this too was theologically motivated) and the expressiveness that the masons who carved the statuary simply could not suppress.

No doubt chemists will go on building their 'beautiful' Platonic molecules. I hope they do. But I can't help wondering what would happen if a truly artistic aesthetic were to become wedded to chemical creativity: if a plastic artist as eclectic as Picasso were to somehow acquire the skills of a Woodward. *That* would be something to see.

Bibliography

Anon. (1964). British report synthesis of vitamin B-12 molecule. *New York Times* International Editon, 6–7 June.

Anon. (1973). Vitamin B_{12} synthesized: a decade-long effort. *Science News* **103**, 22.

Anon. (2003). Primordial recipe: spark and stir. *Astrobiology* 14 May. See http://www.astrobio.net/news/modules.php?op = modload&name = News&file = article&sid = 461.

F. Bacon (1944 [1605/1620]). *Advancement of Learning and Novum Organum*. Willey Book Co., New York.

J. L. Bada & A. Lazcano (2003). Prebiotic soup – revisiting the Miller experiment. *Science* **300**, 745.

J. Baggott (1994). *Perfect Symmetry: The Accidental Discovery of Buckminsterfullerene*. Oxford University Press, Oxford.

P. Ball (1994). *Designing the Molecular World*. Princeton University Press, Princeton.

P. Ball (2001). What a tonic. *Chemistry in Britain*, October, p. 26.

P. Ball (2001). *Bright Earth: The Invention of Colour*. Penguin, London.

P. Ball (2002). Chemistry in soft focus. *Chemistry in Britain*, September, p. 32.

P. Ball (2005). What's in the flask? *Nature* **433**, 17.

N. Bartlett (2003). The noble gases. *Chemical & Engineering News* 8 September, 32.

N. Bartlett (1963). New compounds of noble gases: the fluorides of xenon and radon. *American Scientist* **51**, 114.

N. Bartlett (1962). Xenon hexafluoroplatinate(V) $Xe^+[PtF_6]^-$. *Proceedings of the Chemical Society*, June, 218.

N. Bartlett & D. H. Lohmann (1962). Dioxygenyl hexafluoroplatinate(V) $O_2^+[PtF_6]^-$. *Proceedings of the Chemical Society*, March, 115.

N. Bartlett (ed.). *The Oxidation of Oxygen and Related Chemistry*, p. xi. World Scientific, Singapore.

H. Boynton (ed.) (1948). *The Beginnnings of Modern Science*. Walter J. Black, Roslyn, NY.

W. Brock (1992). *The Fontana History of Chemistry*. Fontana, London.

P. Brook (1990). *The Empty Space*. Penguin, London.

M. W. Browne (1997). Scientists meet analytical challenge of an ephemeral element. *New York Times* science section, 8 July, p. 68.

V. Cohn (1973). Vitamin B-12 project hailed as milestone. *International Herald Tribune* 8 January, 1.

C. Cookson (1981). Synthesis of dodecahedrane. *The Times* 11 February.

R. P. Crease (2003). *The Prism and the Pendulum*. Random House, New York.

A. G. Debus (1978). *Man and Nature in the Renaissance*. Cambridge University Press.

C. Djerassi & R. Hoffmann (2001). *Oxygen*. Wiley-VCH, Weinheim.

A. Donovan (1993). *Antoine Lavoisier*. Cambridge University Press.

P. E. Eaton & T. W. Cole (1964). Cubane. *Journal of the American Chemical Society* **86**, 3157.

P. E. Eaton & R. H. Mueller (1972). The peristylane system. *Journal of the American Chemical Society* **94**, 1014.

R. Eichler *et al.* (2000). Chemical characterization of bohrium (element 107). *Nature* **407**, 63.

A. Eschenmoser (1988). Vitamin B_{12}: experiments concerning the origin of its molecular structure. *Angewandte Chemie International Edition* **27**, 6.

A. Eschenmoser & C. E. Wintner. Robert Burns Woodward: his work on chlorophyll and vitamin B_{12}. Preprint.

A. Eschenmoser & C. E. Wintner (1977). Natural product synthesis and vitamin B_{12}. *Science* **196**, 1410.

G. Farmelo (ed.) (2002). *It Must Be Beautiful*. Granta, London.

L. A. Ford (1993). *Chemical Magic*. Dover, New York.

M. Freemantle (2003). Chemistry at its most beautiful. *Chemical & Engineering News* 25 August, p. 27.

J. C. Gallucci, C. W. Doecke & L. A. Paquette (1986). X-ray structure of the pentagonal dodecahedrane hydrocarbon $(CH)_{20}$. *Journal of the American Chemical Society* **108**, 1343.

J. Hamilton (2002). *Faraday: The Life*. HarperCollins, London.

I. Hargittai (2003). *Candid Science III: More Conversations with Famous Chemists*, p. 29. World Scientific, Singapore.

S. Henahan (1996). From primordial soup to the prebiotic beach. Interview on www.accessexcellence.org. See http://www.accessexcellence.org/WN/NM/miller.html.

J. Henry (2002). *Knowledge is Power*. Icon Books, London.

R. Hoffmann (1995). *The Same and Not the Same*. Columbia University Press, New York.

M. Jacobs (2002). Beauty in the eyes of the beholder. *Chemical & Engineering News* 18 November, p. 5.

M. Jacoby (1997). Heavy elements back on track. *Chemical & Engineering News* July 7, p. 67.

B. Jaffe (1976). *Crucibles: The Story of Chemistry.* Dover, New York.

D. B. Kottler (1978). Pasteur and molecular dissymmetry. *Studies in the History of Biology* **2**, 57.

J. H. Krieger (1973). Vitamin B_{12}: the struggle toward synthesis. *Chemical & Engineering News* 12 March, 16.

H. W. Kroto, J. R. Heath, S. C. O'Brien, R. F. Curl & R. E. Smalley (1985). C_{60}: buckminsterfullerene. *Nature* **318**, 162.

P. Laszlo & G. J. Schrobilgen (1988). One or several pioneers? The discovery of noble-gas compounds. *Angewandte Chemie International Edition* **27**, 479.

A. Lavoisier (1783). *Observations sur la Physique* **23**, 452. See http://web.lemoyne.edu/~giunta/laveau.html.

A. Lazcano & J. L. Bada (2003). The 1953 Stanley L. Miller experiment: fifty years of prebiotic organic chemistry. *Origins of Life and Evolution of the Biosphere* **33**, 235.

H. M. Leicester (1971). *The Historical Background of Chemistry.* Dover, New York.

P. Levi (1986). *The Periodic Table.* Abacus, London.

R. Lougheed (1997). Oddly ordinary seaborgium. *Nature* **388**, 64–65.

G. Maier, S. Pfriem, U. Schäfer & R. Matusch (1978). Tetra-*tert*-butyltetrahedrane. *Angewandte Chemie International Edition* **17**, 520.

G. Maier (1988). Tetrahedrane and cyclobutadiene. *Angewandte Chemie International Edition* **27**, 309.

D. P. Miller (2004). *Discovering Water: James Watt, Henry Cavendish and the Nineteenth-Century 'Water Controversy'.* Ashgate Publishing, Aldershot.

S. Miller (1953). A production of amino acids under possible primitive earth conditions. *Science* **117**, 528.

S. Miller & H. Urey (1959). Organic compound synthesis on the primitive earth. *Science* **130**, 245.

R. P. Multhauf (1993). *The Origins of Chemistry*. Gordon & Breach, Langhorne, Pa.

C. Nicholl (1980). *The Chemical Theatre*. Routledge & Kegan Paul.

K. C. Nicolaou & E. J. Sorensen (1996). *Classics in Total Synthesis*, p.99. VCH, Weinheim.

W. Pagel (1944). The religious and philosophical aspects of van Helmont's science and medicine. *Bulletin of the History of Medicine, supplement 2*.

L. A. Paquette, I. Itoh & W. B. Franham (1975). *meso-* and *dl*-bivalvane (pentasecododecahedrane). Enantiomer recognition during reductive coupling of racemic and chiral 2,3-dihydro- and hexahydro-triquinacen-2-ones. *Journal of the American Chemical Society* **97**, 7280.

L. A. Paquette, D. W. Balogh, R. Usha, D. Kountz & G. G. Christoph (1981). Crystal and molecular structure of a pentagonal dodecahedrane. *Science* **211**, 575.

L. A. Paquette (1982). Dodecahedrane – the chemical transliteration of Plato's universe (a review). *Proceedings of the National Academy of Sciences USA* **79**, 4495.

L. A. Paquette (1984). Plato's solid in a retort: the dodecahedrane story. In *Strategies and Tactics in Organic Synthesis*, p. 175. Academic Press.

L. A. Paquette, R. J. Ternansky, D. W. Balogh & G. Kentgen (1983). Total synthesis of dodecahedrane. *Journal of the American Chemical Society* **105**, 5446.

L. A. Paquette, R. J. Ternansky, D. W. Balogh & W. J. Taylor (1983). Total synthesis of a monosubstituted dodecahedrane. The methyl derivative. *Journal of the American Chemical Society* **105**, 5441.

J. R. Partington (1936). Joan Baptista van Helmont. *Annals of Science* **1**, 359.

C. E. Perrin (1973). Lavoisier, Monge and the synthesis of water, a case of pure coincidence? *British Journal for the History of Science* **6**, 424.

S. Quinn (1995). *Marie Curie: A Life.* Simon & Schuster, New York.

H. S. Redgrove & I. M. L. Redgrove (1922). *Van Helmont.* William Rider, London.

R. Reid (1974). *Marie Curie.* Collins, London.

R. Rhodes (1986). *The Making of the Atomic Bomb.* Simon & Schuster, New York.

D. Riether & J. Mulzer (2003). Total synthesis of cobyric acid: historical development and recent synthetic innovations. *European Journal of Organic Chemistry* **2003**, 30.

D. Rouvray (2005). Exploring the outer reaches. *Chemistry World* June, 42–45.

O. Sacks (2001). Henry Cavendish: an early case of Asperger's syndrome? *Neurology* **57**, 1347.

O. Sacks (2001). *Uncle Tungsten.* Picador, London.

M. Schädel *et al.* (1997). Chemical properties of elements 106 (seaborgium). *Nature* **388**, 55.

M. Schädel (2002). The chemistry of transactinide elements – Experimental achievements and perspectives. *Journal of Nuclear and Radiochemical Sciences* **3**, 113.

M. Schädel (2001). Aqueous chemistry of transactinides. *Radiochimica Acta* **89**, 721.

M. Schädel *et al.* (1997). First aqueous chemistry with seaborgium (element 106). *Radiochimica Acta* **77**, 149.

M. Schädel (1995). Chemistry of the transactinide elements. *Radiochimica Acta* **70/71**, 207.

M. Schädel (2003). The chemistry of superheavy elements. *Acta Physica Polonica B* **34**, 1701.

M. Schädel (ed.) (2003). *The Chemistry of the Superheavy Elements.* Kluwer, Dordrecht.

J. Schummer (2003). Aesthetics of chemical products: materials, molecules, and molecular models. *Hyle* **9**, 73.

J. Schummer (2004). Why do chemists perform experiments? In D. Sobczynska, P. Zeidler & E. Zielonacka-Lis (eds). *Chemistry in the Philosophical Melting Pot*, p. 395. Peter Lang, Frankfurt-am-Main.

G. T. Seaborg & W. D. Loveland. The search for new elements. In *The New Chemistry*, ed. N. Hall, p. 1. Cambridge University Press, Cambridge.

E. L. Smith (1964). A triumph of vitamin chemistry. *New Scientist* 11 June, 660.

G. W. Steeves (1910). *Francis Bacon. A Sketch of His Life, Works and Literary Friends; Chiefly From a Bibliographical Point of View.* Methuen & Co., London.

P. Strathern (2001). *Mendeleyev's Dream.* Penguin, London.

R. J. Ternansky, D. W. Balogh & L. A. Paquette (1982). Dodecahedrane. *Journal of the American Chemical Society* **104**, 4503.

A. S. Travis (1993). *The Rainbow Makers.* Associated University Presses, Cranbury, NJ.

J. Watt (1784). Thoughts on the constituent parts of water and of dephlogisticated air; with an account of some experiments on that subject. *Philosophical Transactions* **74**, 329. See http://web.lemoyne.edu/~giunta/watt.html.

D. Wilson (1983). *Rutherford: Simple Genius*. MIT Press, Cambridge, Ma.

R. B. Woodward, T. Fukunaga & R. C. Kelly (1964). Triquinacene. *Journal of the American Chemical Society* **86**, 3162–3164.

J. H. Wotiz (ed.) (1993). *The Kekulé Riddle*. Glenview Press, Carbondale, Il.

P. Zagorin (1998). *Francis Bacon*. Princeton University Press, Princeton.

Subject Index